N·I·M·B·Y P·O·L·I·T·I·C·S I·N J·A·P·A·N

NIMBY Politics in Japan

ENERGY SITING AND
THE MANAGEMENT
OF ENVIRONMENTAL
CONFLICT

S. Hayden Lesbirel

Cornell University Press | ITHACA AND LONDON

Publication of this book was generously supported by a grant from the Japan Foundation.

Copyright © 1998 by Cornell University

All rights reserved. Except for brief quotations in a review, this book, or parts thereof, must not be reproduced in any form without permission in writing from the publisher. For information, address Cornell University Press, Sage House, 512 East State Street, Ithaca, New York 14850.

First published 1998 by Cornell University Press

Cornell University Press strives to use environmentally responsible suppliers and materials to the fullest extent possible in the publishing of its books. Such materials include vegetable-based, low-VOC inks and acid-free papers that are recycled, totally chlorine-free, or partly composed of nonwood fibers.

Printed in the United States of America

Library of Congress Cataloging-in-Publication Data

Lesbirel, S. Hayden (Sidney Hayden), 1957–
    NIMBY politics in Japan : energy siting and the management of environmental conflict / S. Hayden Lesbirel.
      p.   cm.
    Includes bibliographical references and index.
    ISBN 0–8014–3537–4 (cloth : alk. paper)
      1. Electric power-plants—Location—Japan. 2. NIMBY syndrome—Japan. 3. Energy facilities—Location—Environmental aspects—Japan. 4. Energy facilities—Location—Social aspects—Japan.
    I. Title.
    TK1193.J3L47   1998
    333.793′214′0952—dc21   98–38148

Cloth printing   10 9 8 7 6 5 4 3 2 1

To my mother and to my father

Contents

Maps	ix
Tables	xi
Preface	xiii
1. Conflict, Bargaining, and Compensation	1
2. Project Siting and Compensation	21
3. Structure of the Bargaining Environment	40
4. Gaining Access to Political Power	61
5. Carving Up Opposition Alliances	80
6. Capitalizing on External Shocks	99
7. Dealing with Changing Project Costs	117
8. NIMBY Politics in Japan	135
Notes	155
References	171
Index	183

Maps

Map 1. Electricity spheres in Japan 26
Map 2. Ashihama nuclear plant: site location and property rights ownership 64
Map 3. Hamaoka nuclear plant: site location and property rights ownership 86
Map 4. Matsushima coal-fired plant: site location and property rights ownership 102
Map 5. Tomari nuclear plant: site location and property rights ownership 119

Tables

Table 1.	Characteristics of various energy facility siting disputes in Japan	5
Table 2.	Steps in siting major energy facilities in Japan	24
Table 3.	Power plant lead times in Japan (months)	30
Table 4.	A taxonomy of compensation mechanisms operating in Japan	32
Table 5.	Compensation standards for electric power development	34
Table 6.	Three electric power development laws	36
Table 7.	Explanatory variables and hypotheses with public acceptance times	41
Table 8.	Predicting public acceptance times	44
Table 9.	Evaluating public acceptance times	45
Table 10.	Predictive and explanatory power of the models	50
Table 11.	Key descriptive features of case studies	54

Preface

The NIMBY (not in my backyard) phenomenon has been well documented both in the academic literature and in the popular press. Communities can often delay or force cancelation of proposed facilities they perceive as noxious or ugly, despite the broader social need for such developments. If one reads the Japanese studies literature on environmental conflict, or if one talks to participants involved in the siting, one develops the impression that Japan is no different from other countries when it comes to NIMBY resistance.

Yet while the average time to get local community approval to construct power plants has generally increased since the 1960s, there are significant variations in the times taken to obtain those agreements. Some projects are approved in very short periods of time, others take a very long time, and still others are not approved at all. Such variations are fundamental and cannot be ignored or treated superficially in understanding facility siting.

Notions of Japan as harmonious and centrally controlled have been regarded for quite a long time as an exaggeration. It is now widely recognized that conflict, less bureaucratic control over societal (including regional) interests, and the use of compensation and noneconomic dispute resolution mechanisms form the essence of Japan's consensus politics. Consensual political processes, however, do not yield consistent social choice outcomes. There is much more diversity in policy processes and outcomes in Japan than many existing labels allow us to recognize. I believe that generalizing about diversity is essential to an understanding of Japanese politics.

This book rejects conventional, and often unfounded, views about uniform difficulty in negotiating social agreements in Japan. Conflict is not uniform, and neither are the political outcomes it generates. This book is built on the investigation of one hundred nuclear and conventional power plants developed or abandoned during the postwar era. It explains diversity in siting outcomes by stressing the political economy of dynamic bargaining and creative responses to social problems rather than institutionalized notions of conflict, power, and generic nonresponsiveness. Combined polimetric and case study analyses provide no support for the theory that consensual politics yields consistent siting outcomes.

I set the analysis in a bargaining framework. Bargaining, of course, is not a new concept, and specialists in Japanese studies have stressed the centrality of negotiation in Japan's political economy. This book develops further the notion of bargaining and advances our understanding of conflict, power, and compensation. I treat bargaining not as a cultural process that stresses unique values and conventions. Rather, I treat bargaining as a process of deciding who wins and who loses from social action, and in what forms and by how much losers are compensated. I stress diversity in the intensity of conflict, diversity in the structure of bargaining power, and diversity in negotiating skills. The political economy approach anticipates and allows for diversity among outcomes because of differences in these structural and other factors.

Consensus politics in Japan is about understanding complex and dynamic bargaining. Conflict and bargaining can lead to diverse political outcomes. Negotiated compromises, with innovative use of compensation, are indeed possible and can often happen surprisingly quickly. In short, we need not only more differentiated but also more creatively differentiated models of Japanese politics.

This book is about bargaining and public policy. It explains why siting leads to conflict, how that conflict shapes the bargaining environment, and how negotiators use compensation to resolve environmental disputes. One goal is to refine our understanding about the political economy of public policy in Japan. Another is to further our comparative understanding of siting. Bargaining and compensation are crucial to managing conflict over the siting of noxious facilities not only in Japan but elsewhere as well.

Conducting fieldwork in Japan is not easy. When I entered small Japanese towns and villages, seeking sensitive information on compensation (or who was bought off by whom) from competing groups, it was hard to avoid suspicion, no matter how honestly and openly one operated. At times interviewees were rather unhelpful. One company spokesperson declared, "Even though we have newspaper clippings, there is no point giving them to you, as a foreigner has no hope of understanding Japanese siting issues." On other occasions respondents were extremely helpful. One local mayor said, "I will show you the compensation data," as he proceeded to open a safe, "but you cannot photocopy it." Most interviewees were cooperative and provided me with information.

I could not have completed this book without financial and institutional support. The most important financial support came through the Australia-Japan Research Centre at the Australian National University and then from a Japan Foundation Grant administered by the National University of Singapore. These funds allowed me to work in Japan in the mid-1980s and subsequently with Toyoaki Ikuta and his colleagues at the Institute of Energy Economics in Tokyo. The resource and institutional connections there were crucial to the completion of my fieldwork. Special thanks go to Shinji Suzuki, Mitsuo Takei, Takao Tomitate, and others who, over the years, made my stays so pleasant and productive. I must also mention Hiroko Nishikawa, who provided secretarial and other support far beyond what I could have ever hoped for.

I cannot list all of the people in Japan who assisted me in so many ways. But I express my appreciation to, in alphabetical order, Yoshio Ichiryu (MITI), Takagi Jinsaburo (Citizen's Nuclear Information Center), Teruaki Masumoto (Tokyo Electric Power Company), and Hidenobu Matsuoka (People's Research Institute). A special collective thanks goes to all of the people and organizations at the local and prefectural levels who helped me in Japan.

Feedback from friends and colleagues was extremely rewarding. I am grateful to the following scholars who assisted me in developing and refining my ideas: Lawrence Bacow, Peter Drysdale, Aurelia George-Mulgan, Stuart Harris, Takashi Inoguchi, Ellis S. Krauss, Howard Kunreuther, Margaret A. McKean, Colin McKenzie, Yasuhiro Murota, Michael O'Hare, Ken Oye, Hugh Patrick, T. J. Pempel, Kent E. Portney, Michael Reich, Alan Rix, Richard J. Samuels, Ben Smith, Arthur Stockwin, Nancy Viviani, and John Welfield.

This book has also benefited from the publishing of some of its earlier results. I acknowledge kind permission from the following publishers to reuse some parts of my earlier publications: Institute of Energy Economics, Japan, for my article "Nihon ni okeru hatsudensho no ridotaimu," *Enerugii keizai* 11, 12 (December 1985): 120–137; Kluwer Academic Publishers for "The Political Economy of Project Delay," *Policy Sciences* 20, 2 (1987): 153–

171; University of British Columbia for "Project Implementation in Japan: Origins of Delay in Developing Energy Facilities," *Journal of Business Administration* 17, 1–2 (1987–88): 249–286; Elsevier Science Ltd. for "Implementing Nuclear Energy Policy in Japan: 'Top-down' and 'Bottom-up' Perspectives," *Energy Policy* 18, 3 (1990): 267–282; and Academia Sinica for "Power Plant Siting in Japan: The Role and Effectiveness of Compensation," in *Comparative Analysis of Siting Experience in Asia*, ed. Daigee Shaw, 75–100 (Taiwan, 1996).

Submitting the manuscript to Cornell University Press brought me into contact with Roger Haydon (the editor with a great surname!) who, together with the team there, guided me through the entire process in an open, professional, and encouraging way.

My wife, Edith, and our two children, Ysabel and Adrian, have given me unyielding support in different, yet subtle and powerful, ways in completing this work. The book is dedicated to my mother and to my father, who propelled me to learn about Japan at the age of twelve and, in providing the opportunity for me to go to Japan twice during high school, gave up so much for themselves and my brother, Stephen, and sisters, Annie and Julie.

S. Hayden Lesbirel

Townsville, Australia

N·I·M·B·Y P·O·L·I·T·I·C·S I·N J·A·P·A·N

C·H·A·P·T·E·R O·N·E

Conflict, Bargaining, and Compensation

Delay in implementing a range of private and public projects, including industrial facilities and waste repositories, is an inherent feature of collective action in all nations. Even projects that many regard as desirable, such as hospitals, prisons, and halfway houses, are delayed because of protracted disputes with local communities. Analysts have used various acronyms to refer to the siting problem: NIMBY (not in my backyard), LULU (locally unwanted land uses), BIYBYTIM (better in your backyard than in mine), BANANA (build absolutely nothing at all near anybody), and NIMTOF (not in my term of office).[1]

Such projects are needed because they resolve many societal problems, but their developers have found it increasingly difficult to obtain local community acceptance. Most communities understand the justification for these projects and support them, as long as they are not developed locally. These projects are a bit like weeds and dog droppings: there will always be people who say they are in the wrong place, and the wrong place is usually their own backyard.

This book analyzes the management of environmental conflict over the development of energy facilities. Power plant siting in Japan is essentially a story about fishing cooperatives and private utilities. During the earlier part of this century, Japan established fishing cooperatives and a system of fishing rights to deal with common pool resource problems associated with coastal fishing. During the 1960s and 1970s, Japan created extensive compensation schemes to enable utilities to manage opposition to their facili-

1

ties from fishing cooperatives and other local community groups. Consistent bargaining among utilities, fishing cooperatives, and other local interests, and the use of compensation, are crucial to Japan's history of power plant siting.

There are wide and startling discrepancies in the times required to resolve siting conflicts. For instance, disputes between developers and fishing cooperatives over the Hamaoka nuclear plant took less than two years to resolve, but there was more than a thirteen-year delay before the Tomari nuclear plant proceeded. Conflicts over the Ashihama and Namie nuclear plants have been running for more than thirty years and still are not resolved. Conflict resolution times for conventional projects have ranged from less than two years for the Saijo and Taketomi plants to four years for the Matsushima project and eleven years for the second Gobo plant.

Delay is one indicator of the degree of difficulty in managing disputes over noxious facilities. It reflects the patterns and intensity of conflict, the relative power of interests involved, the effectiveness of conflict resolution mechanisms, and the interactions among these factors in resolving disputes. I ground the analysis of siting conflicts in bargaining theory and use both polimetric and case study techniques to explain variations in negotiating times. I stress the importance of bargaining structures and skill in the use of compensation mechanisms in understanding conflict resolution delays. In doing so, this book refines debates about conflict, power, and compensation.

CHALLENGES AND PUZZLES

Investigating these variations is challenging because they do not appear to have any casually observable patterns, either across different types of technology or space or over time. Disputes over some nuclear plants, such as Tokai and Mihama, took less time to resolve than disputes over some fossil-fueled plants, such as Saijo and Date. Conflict over siting the initial Hamaoka nuclear plant was resolved quickly, but the dispute over siting the Ashihama nuclear plant still continues even though planning for all of these plants started in the 1960s, when there was less community concern about environmental quality. Some conflicts during the 1970s, such as those over the second Hamaoka nuclear plant, were resolved more rapidly than others during the 1960s, such as those over the Tomari nuclear plant. There have been similar variations in conflict resolution times in locating a range of other large-scale projects in Japan.[2]

Most of the available descriptions conclude that developers in Japan face long or indefinite delays in developing noxious projects.[3] Almost all of the projects Margaret McKean discusses were delayed for long periods of time or abandoned. Jack Lewis shows how plans to construct a petrochemical complex in Mishima were abandoned due to strong local opposition. Richard Samuels details the conflict over the Tokyo bridge construction, which continued for more than twenty-five years because implementors could not placate community interests. David Apter and Nagayo Sawa suggest a government failure in developing the Narita International Airport, for which it took more than twenty years to reach a settlement with some local farmers.

The literature has advanced our knowledge of difficult siting conflicts (which is certainly empirically justified for troublesome ones). As significant variations exist, however, our understanding of the management of environmental conflict management will be biased if we focus only on analyzing protracted cases. The existing stock of knowledge can be enhanced further by exploring why the siting of many energy projects is characterized by markedly less conflict and shorter gestation periods. I find it fascinating to explain those "faster-sited" cases and compare them with the "slower-sited" ones.

Power plant siting remains one of the major obstacles facing the Japanese government in achieving its energy policy objectives. The oil crises of the 1970s added impetus to a policy of diversifying away from oil into alternative forms of energy, particularly nuclear energy.[4] This sustained, long-term program sought to provide an insurance cover against further increases in oil prices and unexpected interruptions in oil supplies. During the 1980s and 1990s, Japan has continued to develop nuclear plants steadily despite utilities not opening any green-field sites. A crucial reason for this lies in understanding community addictions to more power plants after the first ones have been built—a "reverse NIMBY syndrome" actually exists for many nuclear plants.

No study of Japanese energy policy can overlook the influential work of Samuels on energy markets, which explores the history of conflict between state and private interests over the ownership of energy-producing facilities.[5] Samuels observes that Japan is one of the few nations where the state has little commercial presence in energy markets. He uses the notion of the "politics of reciprocal consent," a process of permanent negotiation between state and market to decide the terms and conditions under, and the mechanisms through, which control is determined in Japan. He concludes that the private sector maintains control in return for giving the state jurisdiction over markets. Consensual politics in energy markets yields consistent ownership decisions.

Samuels's core finding provides an important entry point for analyzing energy facility siting. Private utilities are responsible for siting and developing power plants in Japan; however, the state wants to ensure that Japan has enough electricity, particularly from nuclear facilities. Government determination to site energy facilities has helped utilities in their siting struggles. The state has not used eminent domain to force the sale of property rights owned by fishing cooperatives but instead has created institutionalized compensation schemes to facilitate bargaining between power companies and those community interests.

This book, by focusing on energy siting, complements Samuels's work on energy policy processes in Japan. It emphasizes negotiations among and within private, state, and regional interests over the allocation of costs and benefits and the role of compensation in policy implementation. Consistent bargaining is as much a feature of decision making as it is a feature of implementation of those decisions in Japan. In terms of ownership, the private sector has had consistently more bargaining power than the state.[6] In terms of siting, however, it has not had similar levels of negotiated control over fishing cooperatives and local communities. Bargaining outcomes can be very different for decision making and implementation even within the same policy arena.

Empirical observations raise three central theoretical issues about facility siting disputes. The first relates to the patterns and intensity of conflicts. The second centers on the structure of power between and within societal interests in different jurisdictions and markets. The third concerns the nature and effectiveness of redistributive mechanisms in resolving siting feuds. I explore each of these issues and their interrelationships briefly as they relate to the Japanese studies and facility siting literatures.

As shown in table 1, this book is primarily comparative in that it contrasts siting disputes that were settled quickly with those that were not. It sets Japanese siting experience in a cross-national perspective, drawing on secondary literature dealing with U.S. and European experiences. The overwhelming bulk of the siting literature is United States–based. Yet this does not imply that the United States is the logical reference point against which always to compare and judge Japan. Among industrialized countries, Japan manages siting conflicts more like Europe than like the United States. I return to this point in concluding the chapter.

Conflict and Resistance

Conflict has been a dominant theme in Japanese society despite the emphasis in harmony models on traditional values, such as deference to authority.[7] Interests in Japan often have heterogeneous and incompatible goals. T. J. Pempel concludes that there are "conflictual outbursts in the midst of consensus" in Japanese society.[8] As Samuels notes so perceptively, "Consensus, even where we find it, is a product of conflict rather than an al-

Table 1. Characteristics of various energy facility siting disputes in Japan

Plant name	Duration	Date range	Plant type	Intensity of opposition	Compensation offered
Mihama	Short	1971–1971	Nuclear, subsequent	Little	Little
Taketomi	Short	1962–1963	Fossil-fuel, initial	Little	Little
Saijo	Short	1967–1968	Fossil-fuel, initial	Little	Little
Tokai	Short	1971–1972	Nuclear, initial	Little	Little
Hamaoka	Short	1967–1969	Nuclear, initial	Little	Medium
Mihama	Medium	1962–1966	Nuclear, initial	Medium	Little
Date	Medium	1970–1973	Fossil-fuel, initial	High	High
Hamaoka	Medium	1972–1975	Nuclear, subsequent	Little	Little
Matsushima	Medium	1973–1977	Fossil-fuel, initial	Little	High
Kashiwazaki	Long	1968–1975	Nuclear, initial	High	High
Gobo	Long	1969–1980	Fossil-fuel, subsequent	Medium	Little
Tomari	Long	1969–1982	Nuclear, initial	Medium	High
Ashihama	Abandoned	1963–	Nuclear, initial	High	Medium
Namie	Abandoned	1967–	Nuclear, initial	High	Little

Note: Qualitative indicators are based on the following criteria.
Duration (months): short (0–24); medium (25–48); long (49–120); abandoned (infinite).
Opposition: little (lack of organized and sustained resistance); medium (some organized resistance initially, but weakening over time); high (considerable organized resistance sustained over time).
Compensation offered (million yen): little (0–4,000); medium (4,001–8,000); high (8,001–15,000).

ternative to it."[9] Conflict occurs in Japan as in other industrialized nations despite a political culture that stresses social harmony and paternalism and a history of repression that makes protest difficult.[10]

Most earlier analyses of environmental conflict in Japan used social grievance models that focused on discontent to explain opposition to noxious projects and delays in conflict resolution.[11] A major step in understanding Japanese environmental conflict was the subsequent use of resource mobilization models. Resistance occurs not only because of social discontent but also because participants can draw on different financial, organizational, and political resources in mobilizing opposition. As Michael Lipsky concludes, "Protest is politics by other means."[12] The Japanese are successful in mobilizing resources to wage protest against noxious facilities both in rural and urban areas, irrespective of the ideology of local elites.[13]

The U.S. literature, often based in collective action theory, argues that hosting hazardous facilities without special compensation arrangements is a "bad bargain" for local communities.[14] The theory assumes that there will be more resistance to than support for noxious facilities and that long siting delays will result. As the benefits of projects are comparatively more diffuse than costs, the higher per-capita cost victims will have an organizational advantage over the many low per-capita win gainers.[15] Many U.S. analysts, like their Japanese counterparts, see community opposition to hazardous projects as perfectly rational in terms of protecting their interests.[16]

Although these various approaches together help to explain intense opposition by fishing cooperatives to nuclear projects at Ashihama and Namie, they do not explain others. Why did the first Hamaoka conflict take significantly less time to resolve given similar levels of community resistance initially? Why did power companies at some stages of the Matsushima and Gobo disputes withdraw their support for projects? Why did the Tomari project take so long to negotiate when there were periods of virtually no conflict between the utility and the fishing groups? Why, despite stronger prefectural support for the Ashihama project than for the second Hamaoka one, did the same utility abandon the former while obtaining approval very quickly for the latter? As some of the comparative literature points out, varying community and developer responses to noxious facilities are not just a Japan-specific phenomenon.[17]

The relationship between conflict and siting outcomes is complex. All social choice processes have conflictual and consensual elements: Conflict is a relative term, and it dominates some social processes more than others. Even if there is considerable resistance, bargaining processes will not necessarily be slow or even result in gridlock. Similarly, even if there is very little opposition, negotiating delays may not be short—delays can also be attributable to developers and community interests not caring very much about whether or not projects are sited.[18] We can understand siting conflicts better by exploring divergences in the value of projects to developers and community interests and by considering two important questions:

What are the origins, patterns, and intensity of resistance and conflict?
What is the nature and strength of the relationship between the intensity of conflict and the difficulty in negotiating social agreements?

Allocation of Power

Decision-making power in Japan has been the subject of many investigations.[19] Of crucial import for this book is the distribution of power between and within local and extra-local, including state and private sector, interests involved in decision-making processes. The most influential argument of the dominant bureaucracy (and one that often has been misrepresented and even ignored) is Chalmers Johnson's work on the developmental state, which focuses on the Ministry of Trade and International Industry (MITI).[20] He recognizes that other interests can influence policy and therefore there is a need for cooperation between state and societal interests. Johnson concludes that the bureaucracy is the dominant player.

The statist paradigm has been challenged widely by both political scientists and economists.[21] Although many see Japan Incorporated, a ruling triumvirate consisting of the bureaucracy, big business, and the Liberal Democratic Party (LDP) as the best explanation of Japanese politics, there is an important set of qualifications to the argument. Pluralists see considerable conflict and competition among the elite. Some suggest that business is more influential.[22] Others argue that the LDP is more dominant.[23] Still others stress the importance of a range of other societal interests in policy making: Michio Muramatsu and Ellis Krauss (patterned pluralism), Takashi Inoguchi (bureaucratic-led, mass-inclusionary pluralism), and Ikuo Kabashima and Jeffrey Broadbent (referent pluralism). Gary Allinson and Yasunori Sone describe how unorganized interests are able to influence policy processes.[24]

The Japanese literature on environmental conflict argues that local communities generally have more power than state and other private interests.[25] Despite a tradition of hierarchical authority over the periphery, local Japanese publics, as Steven Reed points out so well, have considerable autonomy.[26] McKean shows how local protest stifled numerous projects. Lewis details the power of local community interests over very strong commercial and prefectural support. Apter and Sawa show that the use of state force could not easily override local farming and other ideological interests in airport siting. Samuels examines how several communities formed "horizontal coalitions" and stopped in their tracks developers who were seeking to construct a bridge. Even in an unitary state, extra-local governments and commercial interests do not have the ability to implement projects when local publics are hostile to them.[27]

U.S. analyses illuminate how local control over land-use decisions stifles the development of hazardous facilities. Local autonomy is alive and well, and the majority cannot overrule the minority. Some even argue that local communities have too much power. Attempts to increase state power over siting processes through preemption have often failed.[28] Although the state may be able to strip local communities of legal power, it cannot diminish their political power.[29] In contrast, Canadian experience shows that, even in a weak state, there are wide variations in times required to resolve disputes, which is similar to experiences in Japan.[30] The French case is perhaps even more surprising, as variations occur despite highly centralized decision-making structures that coordinate siting.[31]

Bargaining in Japan not only occurs between different levels of government; it also takes place within those jurisdictions. One general conclusion of the Japanese environmental conflict literature is that community interests can often create new mobilizations of power that allow them to dominate electoral processes and policy outputs, a finding mirrored in the U.S. literature.[32] Persons not in elite local decision-making circles have considerable power over siting. One U.S. study claims that because noxious facilities are so unpopular, local elites who support them do so at the risk of their public and political careers.[33] In contrast, many local politicians in Japan and elsewhere actually benefit politically and electorally from actively promoting noxious facilities in their own electorates because of the economic benefits yielded by those projects. One wonders what political and institutional contexts can provide larger political benefits to actions that damage politicians elsewhere.

Whereas local community opposition delayed and killed projects at Tomari and Namie, it did not at other sites. Why could community interests not delay the Mihama and Saijo projects when they had the power to stop them? Why could the prefecture influence the speedy resolution of the Hamaoka dispute but not of the one at Ashihama? Why were utilities in much stronger positions to influence community siting decisions at Mihama relative to Namie? Why did state pressure lead to the utility abandoning the Ashihama project, and why did the lack of any state intervention allow the utility to get agreement quickly for the Taketomi project? Why could local elites manage opposition by fishing cooperatives more effectively at Hamaoka and Mihama than at Tomari and Namie?

The differences in siting outcomes imply that the allocation of power is not always skewed in favor of local community interests. Several observers emphasize the centrality of negotiation, bargaining, and coordination in Japan's consensual politics.[34] Whereas the participants who get involved in siting disputes appear to be stable and predictable, the structure of bargaining relationships and the impact of those relationships on conflict resolution times are not. This book cautions against too much emphasis on the institutionalized weakness or strength of the state. Understanding power requires examination of the complex nature of bargaining between and within different jurisdictional and market levels.

I define power as the ability to get others to do what they would normally not do. It is not *institutional power,* or whether the state or other societal interests are strong or weak, but rather *bargaining power,* which determines social choice outcomes in Japan. Institutional power derives from things such as property rights ownership, dependence on finances, and market structures. Bargaining power derives from what Roger Fisher and William Ury call the "best alternative to a negotiated solution," or the extent to which competing interests have alternatives to reaching a bargain and are willing and capable of pursuing those alternatives.[35] This book distinguishes institutional power from bargaining power and explores a range of issues about bargaining and the management of siting conflicts:

What is the nature and structure of bargaining processes across and within different jurisdictional and market levels?
What determines the bargaining strengths of different interests in conflict resolution?
In what ways does the state influence social choice processes and outcomes at the local level?

Conflict Management

Scattered within the Japan studies literature are observations about the role and effectiveness of compensation in dispute mediation. Several observers treat compensation as a major conflict resolution mechanism in Japan, although many stress other noneconomic instruments, such as privatization of conflict, third-party mediation, and coercion and intimidation.[36] Michael Donnelly argues that side payments are a major conflict management device that can be used as a bargaining chip to accompany a program as a result of a negotiated compromise.[37] Krauss et al. argue that compensation avoids the appearance of zero-sum outcomes and creates the perception of positive-sum outcomes, which help buy off intense opposition.[38]

Many argue that compensation, on balance, does not work well, except under very limited conditions, in resolving conflicts over the siting of unwanted facilities.[39] McKean discusses compensation in passing but implies that it is not effective in managing local protest. Apter and Sawa note how difficult it was for the state to buy off some local farmers at Narita relative to others, without really telling us why.[40] Samuels illustrates that even very innovative forms of political compensation were not adequate to convince the community to agree with the building of the Tokyo bridge.[41]

Despite an earlier belief that compensation was a promising solution to siting problems, much of the subsequent U.S. literature emphasizes difficulties in its use.[42] Michael O'Hare, Debra Sanderson, and Lawrence Bacow argue that compensation agreements are rare because developers find it difficult to identify interest groups whose consent to build projects requires compensation; many of the adverse consequences of projects, such as health effects, are not easily compensated; and it is almost impossible to bind parties to negotiated agreements.[43] Condron and Sipher find that in the United States, states relying on compensation-based schemes have not been any more successful in siting than those attempting to use other, noneconomic arrangements.[44] Kent Portney argues that although risk mitigation (compensation that works on the cost side) produces slightly more change in public perceptions than benefit increases, neither induces shifts in public sentiment to the point where compensation would likely work.[45]

It appears that the tactics and strategies used by opponents often reduce the effectiveness of compensation in resolving NIMBY conflicts. These strategies include forming alliances, influencing electoral outcomes, lobbying politicians, demonstrations and marches, violence, and other pressure tactics. Opponents use such strategies to increase their resources to fight projects and gain legitimacy for their positions. As a political resource, protest then outweighs compensation in the resolution of siting conflicts.

Other works conclude that compensation is likely to be more effective under certain conditions. Susan Pharr emphasizes a strategy of privatizing status-based conflicts, where the state, after resolving initial conflicts, uses preemptive concessions to head off future ones. This implies that subsequent conflicts are easier to resolve after the first ones have been managed.[46] Kent Calder introduces the "crisis and compensation dynamic," a process of accommodation in which government creates circles of compensation (providing benefits through policy and other rewards) to opponents during perceived political and economic crises in return for their political support.[47] He suggests that compensation works better in resolving conflicts during crisis than in periods of calm. Howard Kunreuther and Douglas Easterling, in an analysis of U.S. experience, conclude that compensation is not likely to work unless some perceived threshold level of safety is assured and trust can be built during negotiations.[48]

These contentions do not hold up consistently against the evidence. Why did compensation not work in ameliorating fishing opposition at Ashihama when supporters were willing to pay much more than at Hamaoka, where safety concerns were higher? Why did the utility pay significantly more to fisheries at Matsushima relative to the initial Hamaoka case, yet the former dispute took longer to resolve than the latter? Why was the power company able to pay little compensation at Mihama and obtain a very quick settlement? Why did compensation not work in resolving the Namie dispute, which occurred during a crisis period? Why did the utility pay compensation to Kyowa village when it abandoned its site location there?

This book adds to knowledge of strategies for managing siting conflicts.[49] Compensation mechanisms are central to resolving disputes over environmentally risky facilities in Japan and elsewhere, and negotiated compensation agreements are indeed possible. The effectiveness of compensation will, however, depend in an important way on the structure of bargaining power and bargaining relationships, which raises several issues:

What is the nature and role of compensation mechanisms in conflict management?
What is the relationship between compensation and other noneconomic dispute management strategies?
Under what conditions are compensation mechanisms likely to be effective in resolving disputes?

Both the Japanese and the U.S. siting literatures suggest that siting conflicts are protracted because communities believe that noxious facilities are not in their interests; because they have or can acquire political power; and because conflict resolution mechanisms, including compensation, rarely work well. Evidence from Japanese siting experience points to less stability in these interactions and their impact on conflict resolution times. For instance, developers experienced roughly equal delays for the Matsushima and Date projects when, at the former, there was very little resistance, more state intervention, and less compensation offered, compared to the latter. This book sheds light on these inconsistencies and provides a more comprehensive explanation of the management of siting conflicts.

A BARGAINING FRAMEWORK

In bargaining, actors generally start with and continue to advance widely differing positions and strategies. Bargaining on a regular basis, however, tends to move actors through making compromises toward common ground. Negotiating can prevent deadlocks, even if high levels of conflict exist initially. Bargaining encourages participants to find innovative and creative solutions to social problems. Compensation is an important mechanism that facilitates compromise about burden sharing among conflicting parties. Bargaining and compensation influence how long it takes to site hazardous facilities.

Bargaining is highly complex and takes place interactively among parties both within and across inter- and intrajurisdictional and different market levels. Participants use multiple strategies to enhance their positions depending on the bargaining environments. They often face high levels of uncertainty in deciding how best to approach negotiations. The processes are dynamic, and players constantly must adapt to changing situations that influence their positions and their strategies. Participants make mistakes—often bad ones—but they also can learn from them.

A bargaining framework, emphasizing the concepts of conflict, power, and compensation, provides the theoretical frame of reference for analysis. Conflict over facility siting arises when there is a divergence in the value (costs and benefits) of projects to developers and community interests. Three levels of "community" are relevant to the analysis. "Local community" relates to jurisdictional boundaries, such as villages or cities, where projects are developed. "Regional community" is defined to include the relevant local community and neighboring local communities that are subject, for whatever reason, to negotiating processes. "Prefectural community" refers to the prefecture in which the project is situated.

Private or public developers generally develop projects to achieve, among other things, market and security objectives. Projects impose externalities, or an array of both positive and negative spillover effects, on community interests. These spillover effects can be both monetary and nonmonetary in nature. There may be financial, employment, and other social gains that result from hosting projects. There may be costs, such as health risks, property value losses, and electoral costs.[50] Resistance to projects occurs when communities judge that net spillover effects are negative or when they identify inequality between who gets the benefits and who has to accept the adverse impacts and risks.

When such divergences exist, and when communities have veto power through legally recognized property or zoning rights, project siting will require negotiation. This process is characterized by bargaining between developers and community interests over the allocation of costs and benefits expected to be generated by the development. Those who are affected or expect to be affected may attempt to use a range of strategies to increase their share of the spoils.[51] Bargaining is essentially concerned with who gains and who loses and to what extent and in what forms losers are compensated.[52] To achieve agreement, developers will have to redistribute some of their expected gains to communities expected to be affected adversely.

The major mechanism effecting this redistribution is compensation. The theory of compensation posits that peoples' perceptions of facilities are related to their benefit-risk calculations. Various types of compensation, such as side payments, risk mitigation, and the provision of public goods, are aimed at making people at least as well off after the project is sited as they were before.[53] Compensation aims to redress the disproportionate allocation of costs and benefits between communities accepting projects and the broader society. Strategically and politically, it realigns the interests of critical actors with those of the public interest.[54] By increasing the benefits, or at least reducing the costs, of projects to affected parties, compensation reduces or nullifies resistance and facilitates bargaining. Compensation seeks to produce a positive-sum result, with local communities, developers, and the national public all benefiting from facility siting.[55]

The resolution of siting disputes requires that developers and community interests strike an acceptable bargain over expected gains and losses from projects. An acceptable bargain is one that is politically and economically tolerable to all parties given the allocation of bargaining power. The availability of alternatives (opportunity costs, including social and political costs, of alternative courses of action) will set the limits within which bargains are conceivable. Hence, the limits will be determined by the price at which both the promoter and the community would see the alternative option of not striking a deal as being less costly or more attractive.

Whether an acceptable bargain can be reached will depend on the total compensation that can be paid to community groups, as well as the distribution of that compensation within the community. A necessary condition for striking an acceptable bargaining is that it falls within bargaining limits. For example, an acceptable agreement will not be reached when a community has very viable options to accepting a facility and demands compensation to the point at which a promoter prefers to develop a project at an alternative location. The sufficient condition for an acceptable agreement is that the distribution of compensation is consistent with the relative bargaining power of different interests. For example, an acceptable agreement will not be forthcoming when most of the available compensation is paid to some community groups at the expense of other groups that have or can acquire veto power.

The speed at which projects are approved and their final form will be related to the extent to which redistributive mechanisms operate effectively. Chapter 2 provides evidence that the mere existence of these arrangements does not guarantee stability in bargaining times. Compensation mechanisms are likely to be more effective when there are large expected net benefits from projects (or large economic surpluses available for redistribution) and when compensation can be used to offset the expected adverse effects on community interests given the structure of bargaining relationships. Six factors are likely to influence the effectiveness of compensation in resolving facility siting conflicts.

Economic Surplus Derived from Net Benefits and Costs. The expected economic surplus from projects conditions the broad structure of the bargaining environment in which negotiations are conducted. The surplus will be determined by the value that promoters and communities attach to facility siting. Benefits and costs are not likely to be valued in the same way by all interests in all situations: A poor fisherman is more likely than a rich one to value a marginal 1,000 yen more highly.[56] The extent to which expected benefits exceed costs influences the economic surplus available for redistribution to win community approval. Resistance is likely to be stronger when large adverse effects are expected from projects and when there are difficulties in compensating for negative spillover effects. Under those circumstances, the bargaining environment is likely to be structured to impede conflict resolution.

Distribution of Benefits and Costs. The allocation of costs and benefits among players may vary, which is likely to affect bargaining outcomes. Some projects may appear to impose large concentrated per-capita costs on organized interest groups, defined in functional terms as "private organizations that have formal organizational charters and that seek the promotion and protection of the collective interests of their members."[57] Others may seem to impose costs on disparate and unorganized groupings. Even if net expected benefits are the same, there is likely to be more resistance if organized groups expect large costs. Those affected are likely to sustain opposition effectively as the per-capita costs of not doing so will be relatively high, thereby prolonging bargaining.

Bargaining Power Derived from Alternatives. The allocation and use of bargaining power are also likely to influence the effectiveness of compensation in resolving siting conflicts. The bargaining power of parties will depend on how attractive to each is the option of not reaching an agreement.[58] The extent to which developers and community interests have less costly (or more beneficial) alternatives will influence the willingness to compensate or be compensated. At any time, the bargaining situation may favor certain interests over others. Siting is likely to be more difficult when local communities have viable alternatives for social and economic development. They will be in stronger bargaining positions than even large private companies as they will be able to drive up the marginal negotiating costs to the point at which developers prefer to develop other projects.

Bargaining Skills. The effectiveness of compensation may also be determined by the skill with which players use negotiating strategies. Developers and opponents generally aim to alter the expected impacts of projects, the distribution of power, or both in order to change the bargaining environments. Such strategies may include scheduling of negotiations, third-party intervention, alliance formation, the supply of information, project design modification, and the use of force and pressure. Developers are more likely to be able to resolve conflicts if they can devise strategies that legitimately create a negotiating table on which bargains about compensation and other requirements can be reached.

Changing Expectations about Bargaining Positions. Changes in expectations or changes in perceptions about future social and political orders or states of the world can inject instabilities into bargaining. These instabilities can occur endogenously due to such things as local electoral outcomes, changes in strategic alignments, and choice of strategies, which are under the control of participants. Instabilities can also occur exogenously due to such things as national policy and market changes, outcomes at other sites, and international events, which are not under the control of participants. Such changes can affect bargaining processes and, depending on their direction and strength, can alter the responses of interests and the structure of the bargaining environment. Conflict resolution is likely to be easier when changing expectations increase the benefits of projects, thus allowing promoters to manage competing compensation claims.

Uncertainty about the Bargaining Environment. Complete knowledge about future costs and benefits or social and political orders never prevails in bargaining. Interest groups will be uncertain about the nature and extent of spillover effects or the implications of different negotiating strategies. For example, community interests may not have the information or capability to assess the impacts of projects on them. The extent of the uncertainty influences the expected economic surplus that is available for redistribution to resolve disputes because reducing uncertainty involves some cost. Developers facing high levels of uncertainty may not be prepared to pay as much compensation to offset uncertain project costs. Under these conditions, compensation requirements are likely to be more difficult to manage, and settlements may take longer.

In summary, compensation mechanisms are likely to be more effective in resolving siting conflicts, and negotiating delays are likely to be shorter where:

The bargaining environment is structured to yield large expected economic surpluses from projects;

The expected distribution of costs falls unevenly across disparate or unorganized groups opposing projects;

Communities are vested with relatively less bargaining power than developers;

Developers can execute strategies that legitimately create a negotiating table on which competing compensation claims can be managed;

Changing expectations increase the net benefits of projects and allow developers to deal with compensation requirements; and

Uncertainty about the bargaining environment is low or can be handled effectively.

The factors that influence bargaining times interact in complex ways. For example, even though the structure of the bargaining environment may be exactly the same in two disputes, settlement times may differ because of the way promoter strategies interact with the overall structure of the bargaining environment. In one case, local community interests may regard promoter strategies as being legitimate, which may interact with the bargaining environment to facilitate a settlement. In another case, community interests may perceive that promoter strategies are unacceptable, which may intensify resistance, making the bargaining environment less conducive to an agreement. A multivariate examination of conflict resolution helps illuminate these interrelationships and their impacts on bargaining times.

Chapter 3 develops a polimetric model for analyzing variations in conflict resolution times. It regresses a set of factors that are likely to condition the structure of the bargaining environment on a lead time variable, specified and measured in Chapter 2. These factors include the market need for projects, the extent of social and economic opportunities, the size of the rural sector, the political and ideological orientation of the community, the risks attached to different technologies, and the social attitudes toward environmental preservation. For example, expected supply shortages can be taken as one measure of the market need for projects. Developers are likely to have a stronger incentive to develop facilities where expected shortages are relatively high. Even if all other factors were equal, developers who attach a strong weighting to the benefits of projects are likely to be more willing to provide adequate compensation to resolve conflicts.

This regression approach examines the overall effect of these factors, often working in different directions, on the times required to resolve environmental conflicts. It provides a quantitative measure of the relative significance of the structure of the negotiating environment, relative to other influences, on complex bargaining outcomes. The approach considers the degree to which approval times can be predicted at the beginning of bargaining and whether any predictability is contingent on changes that emerge during disputes. It also assesses the possibility of identifying a set of more global, as opposed to more site-specific or idiosyncratic, determinants of bargaining times.

Understanding the management of environmental conflict does not end with generating polimetric models. Regressions measure residuals, the difference between observed and predicted times, and identify disputes when the bargaining environment is a reasonable explanator of delay and when it is not. Quantitative approaches make assumptions about the measurement and distribution of costs and benefits, the allocation of bargaining power, the use of bargaining strategies, changing expectations, and uncertainty, which may not be captured fully or even partially in any statistical analysis. Furthermore, quantitative approaches do not reveal very much about the nature and structure of dynamic interactions between quantitative and qualitative variables and how those interrelationships influence conflict resolution. A combination of polimetric and case study approaches can enhance understanding of conflict resolution in ways that would not be possible by relying on one approach or another.

Chapters 4 to 7 explore through the use of narratives the unexplained variation in settlement times and provide further insights into facility siting processes in Japan. I selected the cases on the basis of the error analysis and their representativeness of siting conflicts. Chapter 4 examines the Ashihama nuclear case. Local opposition was able to gain access to political power and force abandonment of the project. Chapter 5 explores the Hamaoka nuclear dispute. Regional power brokers were able to split an economic and ideological alliance and win agreement very quickly. Chapter 6 considers the Matsushima coal-fired conflict. The developer took some time dealing with another utility and the bureaucracy before completing community negotiations. Chapter 7 examines the Tomari nuclear dispute. Increasing management costs of a project delayed negotiations significantly until the utility finally received agreement by changing the location of the project.

The empirical database on which the analysis rests relied on creating several new data sets. Information on power plant lead times was gathered through an extensive survey of Japan's power companies. Other data were assembled from utility documents and other primary sources, local government surveys, regional newspapers and other secondary sources, and extensive interviews with players from all sides of siting disputes at different jurisdictional levels.

The bulk of the analysis covers the period from the early 1960s to the early 1980s. Siting experience during this period matters in understanding more contemporary developments. Since the mid-1980s, Japanese utilities have not garnered community agreement to develop any green-field nuclear sites. Power companies have been able to site nuclear capacity only at locations opened previously. Unless power companies can get community consent at new green-field sites, nuclear growth will come to a halt once existing sites are exhausted. Several local communities in the 1990s have decided to hold community referenda to vote on accepting nuclear plants. As the recent Maki plebiscite, which rejected a nuclear plant, shows, these referenda pose a major challenge to the nuclear industry as it enters the next century.

IMPLICATIONS

This book uses evidence from energy siting in Japan to test its propositions. The results are important for understanding Japanese politics and facility siting more generally. One aim is to relate the findings to some broader contentions found in the Japanese politics literature. Although principally focusing on Japanese details, another aim is to provide broader theoretical and comparative insights into facility siting.

Japanese Politics

Most of the more general analyses of Japanese politics focus on the distribution of power between different interests at the national level. While that is clearly crucial, the analyses shed little light on the importance of subnational interests and their implications for political power in Japan. Where they do, regional politics are assumed to matter very little in the broader allocation of power in Japan. The reality is different. As the local government and environmental conflict literatures illustrate, when policy adversely affects local and regional interest groups, subnational governments will matter. How and when these governments can exert power must be integrated into more general treatments of Japanese politics. Siting is a prime example of subnational politics having a crucial bearing on the achievement of national policy objectives.

Many observers tend to treat social choice outcomes in Japan as uniform, although the policy literature collectively shows diversity in policy processes.[59] They see Japan as looking like this or like that in their attempts to generalize about the Japanese politics.[60] Generalizing is important, but it needs to account for case-to-case variations when they occur both across and within policy arenas. It is not difficult to find diversity in Japan and, when we discover that it is meaningful, our more general frameworks should explain its implications. The experience of project siting shows that it is equally possible to generalize about diversity and the reasons for it. Where we begin and where we arrive are as much empirical questions as analytic ones.

An often-heard observation about Japan is that it is a generic nonresponder.[61] Leaders rarely respond to change and have fixed solutions to most similar political problems. For better or for worse, politics always looks pretty much the same, at least within any single policy arena. Many observers argue that Japan responds to change only with external (and this often means U.S.) political pressure, and even then, it does not change that much. Siting experience shows that these conclusions are the result of an oversimplifications. Japan's regional political and corporate leaders do respond to changed conditions and often with great skill because it is in their interests to do so to manage social problems. Bargaining in Japan does lead to creative solutions and very different political outcomes.[62] I stress the importance of not only more differentiated but also more creatively differentiated models of Japanese consensual politics.

Facility Siting

This book is also relevant in developing more comprehensive project evaluation methods, most of which use Cost-Benefit Analysis (CBA). Least-cost techniques specify and then aggregate all relevant benefits and costs associated with project development. They suggest a project should proceed on efficiency grounds, providing a potential Pareto improvement exists. In this context, there is an improvement if aggregate discounted benefits exceed costs and individuals disadvantaged by the project could, in principle, be compensated by those who were advantaged and still remain at least as well off as before the change.[63] These approaches yield a theoretical ordering of project development—projects being scheduled in accordance with net benefit-cost ratios.

A standard treatment of CBA is useful for selecting high-return projects independent of bargaining and compensation requirements. Although Japanese utilities use least-cost approaches in site selection, they also use compensation in obtaining community approval for projects. Yet there are still often large discrepancies between this theoretical ordering of sites and the actual ordering of when agreements were reached. In Japan, winning approval for many projects slated for early development in the 1960s actually took longer than many proposed in the 1980s. Least-cost projects, in terms of economic efficiency, can be among the highest-cost projects, in political terms, even if compensation mechanisms exist and compensation is paid. Project planners must consider explicitly the bargaining and compensation costs in reaching agreements and take them into account in evaluating alternative projects.

There is still a debate about the role of culture in Japanese politics. Culturalists suggest that country-specific models should be used to understand bargaining because Japan's social conventions, practices, and mores are so different from those of the rest of the world. Others stress the importance of institutional and structural determinants. Evidence from siting experience supports the latter contention. We can explain siting outcomes in Japan, in both urban and rural areas, by grounding our analysis in bargaining theory. Cultural explanations are not necessary to account for observed variations, even assuming some cultural diversity at the regional level.

The overwhelming bulk of the siting literature focuses on U.S. experience. The United States is different comparatively in terms of energy siting. Delays have been consistently longer in the United States, particularly since the 1970s, than in Japan and European nations such as France, Germany, and Spain.[64] Some projects have been delayed significantly and even abandoned in Japan and Europe, but many have not. Unlike the United States, these countries have been able to use compensation to resolve many siting conflicts. Siting processes certainly do not always end up in gridlock. Negotiated outcomes with the use of compensation happen quite often.

The siting literature has not analyzed the variations in times required to get political agreements for siting noxious facilities in the ways this book does, and a systematic cross-national investigation is not possible. Japan's rich experience in energy siting does, however, enhance our understanding of the importance of bargaining and compensation in the management of NIMBY disputes. I hope that this book stimulates further comparative research in this area of enquiry that is becoming more important for all nations.

C·H·A·P·T·E·R   T·W·O

Project Siting and Compensation

Siting power plants in Japan and other advanced countries requires a planning stage and an implementation stage, which consists of public acceptance, licensing, and construction. Total lead times required to develop energy plants in Japan are highly variable and have become on average longer since the 1960s. At some stages of siting there is more stability; at others there is less. The major source of variation in total lead times across space and over time is the instability in public acceptance times. Political bargaining processes at the subnational level have a major effect on the speed at which Japan achieves its energy supply objectives.

Japan, like Europe, has created a sophisticated set of compensation mechanisms for managing environmental spillover effects associated with siting large-scale facilities. It developed these mechanisms in response to the establishment of agricultural and fishing cooperatives to manage common pool resource problems. These arrangements offer developers a wide range of legally acceptable ways to negotiate siting agreements with property rights holders and other community interests. They work well in resolving some siting conflicts but not in resolving others. The mere use of institutionalized compensation arrangements does not guarantee stability in settlement times.

PLANNING AND IMPLEMENTATION

In contrast to most other advanced nations, where, according to Samuels, "state-owned monopolies are the usual form of organization for the [electric power] industry," nine private utilities own most of the electric-generating capacity in Japan.[1] Only limited electricity has been generated by the Electric Power Development Company (EPDC), in-house generation by large firms, and prefectural governments. Samuels's finding that the private sector controls electricity markets is important in energy project planning and implementation.[2] Although the Ministry of International Trade and Industry (MITI) has legal jurisdiction over markets, private utilities—not public authorities—plan and develop energy facilities in Japan.

One factor motivating utilities to site new facilities is their expected electricity supply and demand. Strong incentives encourage power companies to supply adequate electricity to grids. Article 2(b) of the Electric Utility Industry Law legally obliges utilities to supply electricity to consumers. They face large financial penalties for failing to do so.[3] Utilities annually prepare medium- to long-term predictions of electricity demand. A risk component is added to this forecast demand to cover unexpected shortages from plant malfunctions or from unanticipated demand increases. Installed capacity and capacity that will be decommissioned is calculated.[4] When supply and demand forecasts point to the need for more electricity, utilities identify sites for additional projects.

Site selection is a continuous task and involves using least-cost techniques to establish a pool of candidate sites. There are five major technical criteria for inclusion: the existence of flat and stable terrain, the availability of cooling water, a relatively low population density (particularly for nuclear plants), accessibility to transportation routes, and proximity to major load centers.[5] The technical decision to convey candidacy status to a particular site is made by attaching relative weights to selection criteria. For example, when the features of two sites are similar except for proximity to load centers, the closer site will be included because it will be cheaper to construct transmission lines.

Although these five criteria are universal, there are three important locational features of siting in Japan.[6] The first is that, apart from hydroelectric plants, almost all of Japan's nuclear and fossil-fueled plants are found in coastal regions. Mountainous inland regions offer little flat and stable terrain, and the shallow, fast-flowing rivers cannot provide sufficient cooling water. Inland location is also impractical because of difficulties in transporting large prefabricated construction components. In contrast to Japan, many nuclear and conventional power stations are located in inland America and in some European countries.[7] As Japanese utilities look to the coast for plant sites, they have to negotiate with fishing and agricultural cooperatives.

The second characteristic siting feature is the relatively high concentration of plants in a given area. The average concentration in European countries is two to three plants per site, but it is not uncommon for Japanese utilities to locate four to eight plants at a single site. Site availability in Japan is limited because of climate, scarce land, competing claims for resource use, and other historical reasons. As one MITI official noted, "Even though Japan's coastline is actually longer than America's, we are really constrained in siting. Locations are out in north Hokkaido because the seas are frozen for most of the year and do not provide sufficient cooling water. Utilities would be committing corporate suicide proposing nuclear plants in Hiroshima or Nagasaki prefectures, for obvious reasons."[8] Developers want to make optimal use of scarce sites but, as we shall see, they also concentrate plants in specific locations for political reasons.

The third siting feature is the ruralization of power plant location. Over time Japanese utilities have used up scarce urban sites and have had to site both nuclear and conventional plants increasingly farther away from these urban locations. Average distances from major load centers grew more than twofold from the 1970s (Tokyo: seventy-two kilometers and Osaka: forty-six kilometers) to the 1980s (Tokyo: one hundred eighty kilometers and Osaka: ninety-nine kilometers). Utilities are increasingly reliant on coastal sites in more rural areas for siting power stations.

After deciding to develop additional capacity, a power company will choose a site or sites from its candidate pool. It will then obtain community acceptance, receive necessary permits, and construct the project. Key starting and finishing points for each stage are defined as veto points, or decisions that must be made by relevant participants to enable a process to start or to proceed to a subsequent stage. Veto points are critical as projects may be delayed or abandoned at any one of them. Table 2 presents a taxonomy of the major steps together with significant decision points, criteria for reaching them, and key actors who judge whether criteria have been met. Comparative analysis suggests that other industrialized nations follow a similar sequence in siting.[9]

These veto points were derived from a survey of Japanese utilities. Company documents and extensive interviews with promoters and opponents of power plants helped to design a questionnaire. I sought information on the dates critical decisions were made to provide some measure of the time involved in each stage of siting. All of the utilities responded, and ninety percent of total questionnaires returned (ninety cases) were used in the analysis.[10] The examination covered the period from 1960 to 1985. As relatively few power plants have been sited since 1985, the

Table 2. Steps in siting major energy facilities in Japan

Step	Process	Key veto points	Key criteria for veto points	Key interests judging criteria
Public acceptance	Bargaining over burden sharing	Utility decision Declaration of interest Application to site Local invitation	Technical criteria Sociopolitical assessments	Utilities Subnational government or elites
Licensing	Regulation to balance risks and benefits	EPDCC approval	Subnational approval Prospects for rights transfer Impact assessment Public hearings	Utilities Property rights holders Subnational and national governments
Construction	Optimizing construction costs	Construction Planning Permit	Completion of rights transfer Necessary licenses	Subnational and national governments
Operation	Optimizing operating costs	Commercial Operating Permit	Construction of plants	Utilities and plant makers Subnational and national governments

Sources: Shigen enerugii chō (1985), internal company documents, interviews, and survey data.

study provides a comprehensive treatment of Japanese energy siting experience.

It is impossible analytically to separate out different stages of siting. Stages do overlap. Public acceptance can be problematic during licensing, construction, and even operation and decommissioning.[11] Some licences have to be issued during public acceptance and construction. The markers for ending different stages can be defined as follows: the Electric Power Development Coordination Council (EPDCC) approval as the end of public acceptance, the Construction Planning Permit (CPP) as the end of licensing, and the Commercial Operating Permit (COP) as the end of construction. In terms of process, the most important steps are generally completed by the time these markers are reached. The quantitative results provide judgment as to how meaningful any overlap is in the context of the analysis.

The site selection process culminates in a decision to develop a power plant. This decision can take various forms that generally include a decision at an internal board of directors meeting, a declaration of intent by a utility to regional government, an application for an investigation permit, the appearance of a particular location in a utility's construction plan or an invitation by a local government.[12] Any one or combination of these events usually indicates that a utility has selected a particular location on which to site a project.

The *public acceptance stage* is defined as the period from a power company's decision to build a facility at a specific location to EPDCC approval. It involves reaching broad political agreement over the proposed power plant, which is accomplished through negotiations between and among promoters and community interests in different jurisdictions, and is concerned with how expected project costs and benefits will be ultimately shared.

The EPDCC, under the jurisdiction of the Economic Planning Agency (EPA), considers siting proposals. Membership comprises the prime minister, ministers and directors from various ministries and agencies, bureaucrats, businesspersons, and other learned members of society. This approval signifies national agreement on market need for power plants and broad political consent for the project at the subnational level.[13] Siting can proceed to the substantive part of the licensing stage only after EPDCC approval.

An important criterion for approval is agreement between MITI and utilities on the need for a project to balance electricity markets. As shown in map 1, utilities enjoy monopoly control over supply to specified electricity spheres. For example, Tokyo Electric supplies electricity exclusively to the Kanto electricity sphere. From a perspective of national electricity management, these utility spheres are aggregated into three regional electricity spheres: eastern, central, and western.[14] Power companies are primarily interested in balancing electricity markets in more narrowly defined utility spheres. MITI is more concerned with markets in broader regional spheres. Discrepancies between power company and broader regional requirements for incremental capacity can influence siting processes.

Agreement from local and prefectural governments is a prerequisite for EPDCC approval. This approval is influenced by the extent of broad regional acquiescence to the project. Approval does not require the completion of formal property rights transfer arrangements. It only requires local mayors and prefectural governors to make political judgments that in-principle agreement has been reached for property rights transfers and that there is

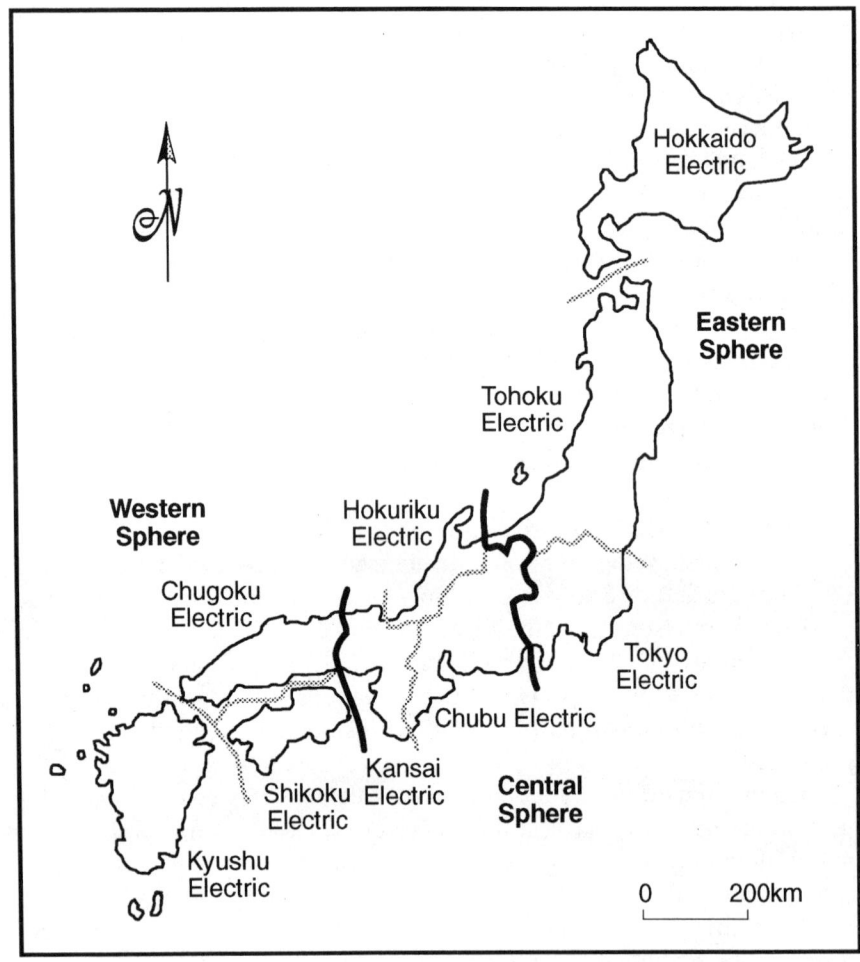

Map 1 Electricity spheres in Japan

a favorable outlook for eventual settlement of those transfers. There are no explicit criteria that require community compensation arrangements to be finalized before EPDCC approval is granted.[15] Acceptance by local authorities may, however, be influenced by community demands that compensation be paid in return for giving consent to a project. Since 1974, approval also requires a public hearing based on an Environmental Impact Assessment. Developers prepare a statement, and MITI conducts hearings to allow communities to examine these reports and voice their opinions.[16]

The *licensing stage* is the period from EPDCC approval to the issue of the CPP. The regulatory process involves the government using its licensing power to strike its preferred balance between the interests of the national community in terms of electricity supply and economic security and the interests of subnational communities in terms of the environmental and other risks. Construction proper can commence only after MITI has issued the CPP.

The CPP denotes that agreements on property rights transfer and other community arrangements have been completed and that all relevant licences have been issued.[17] Approval by EPDCC requires only in-principle agreement on property rights transfer. In many cases negotiations continue throughout the licensing stage, but negotiators are generally concerned with final marginal compromises. The CPP requires, however, that all relevant property rights have been formally transferred to utilities.

About fifty to sixty permits and licences are issued during the licensing stage, some by local governments but most by prefectural and national governments.[18] Some permits, such as those relating to preliminary investigations and use of roads, are always required, irrespective of the location or type of plant. The need for other permits depends largely on site-specific and technological considerations. For instance, proposals to build power stations in or near national parks require permits under the National Park Law, whereas proposals to develop nuclear plants necessitate the Nuclear Reactor Permit.

The *construction stage* is the period from the issue of the CPP to the granting of the COP. Construction is an economic optimization process. Plant makers, such as Toshiba and Hitachi, under contracts from power companies, try to minimize construction costs subject to resource input (labor and capital), technical constraints, and the urgency of delivering power to the grid.

The construction stage is subdivided into preliminary construction and construction proper.[19] Preliminary construction involves clearing the site and providing infrastructure necessary to build and operate plants. It can start before the issue of the CPP, providing property rights transfer arrangements have been completed and local authorities have given their consent for preliminary construction. Construction proper commences with the excavation of the land on which the boiler (conventional) or reactor (nuclear) is to be placed. This can take place only after MITI issues the CPP.

In summary, there are four important characteristics of energy siting in Japan. Private utilities—not state-owned corporations—are the major planners and implementors. Site location is almost always confined to coastal areas, and there is a heavy concentration of plants in those areas. Coastal location means that utilities have to deal with both fishing and land rights holders in bargaining. Utilities are becoming more reliant on rural sites, increasingly farther away from urban areas.

THE MEASUREMENT OF LEAD TIMES

Identifying sources of variation in total lead times requires an analysis of average lead times and their coefficients of variation (COV) for each stage of project development. The COV indicates the degree of variance around an average measure. For example, a relatively small COV for a particular stage suggests relative stability in times for that stage. It is possible to assess the stages that contribute most to total lead time variation by using averages and COVs. A stage with a relatively long average time and a relatively high COV will contribute more to total variation than one with a shorter average time but a similarly high COV.

The analysis divides power plants into *nuclear* (light and heavy water reactors) and *conventional* (oil, coal, and gas) plants. It does not consider facilities as *individual units*, but as *packages*, each package consisting of one or more plants. Utilities generally purchase property rights and reach community agreement to develop several power stations, rather than proceeding on a sequential plant-by-plant basis. They often develop more than one package at any given site. An *initial package* consists of plants sited at a location where there are no existing plants. A *subsequent package* consists of plants located at a site where facilities are being licensed, under construction, or operating. Measures refer to average times for the given package.[20]

There are two conceptual and measurement problems in measuring lead times. First, there are a number of decisions, such as a power company declaration or local government invitation, that identify the beginning of public acceptance. Examination of the data suggests that decisions that indicate starting points vary across the sample; however, differences between the times registered for any combination of these decisions are normally one to two months, with the largest being four months. As such differences matter very little, the earliest time registered in each case is used.[21]

Second, EPDCC approval does not necessarily imply that all relevant bargaining processes are complete. For example, final marginal negotiations over property rights transfers can influence licensing. There is fuzziness between different stages of project siting. The EPDCC is, however, a useful marker for declaring that key bargaining, in relation to criteria for its approval, has been sorted out and that the major part of licensing can commence. As there is much broader scope for public acceptance to influence licensing, relative to construction, it is useful to consider preconstruction lead times (defined as public acceptance plus licensing lead times) before exploring those two component stages separately.

Table 3 presents major patterns of lead-time behavior. Average total lead times for nuclear plants (154 months) is roughly twice as long as for fossil-fueled plants (75 months). The same general pattern is observable in both preconstruction and construction times. The F-test provides evidence that these differences are statistically significant. The licensing stage, which can be highly politicized, appears to contribute relatively little to total lead times for nuclear (22 months) and fossil-fueled (12 months) plants. Public acceptance and construction lead times are much longer and account for a high proportion of total average lead times. Despite the analytic problem of overlap, licensing times do not greatly affect total lead times.

An interesting exception was the Onagawa nuclear dispute, resolved during the 1980s. Soon after EPDCC approval but before property rights transfers negotiations were finalized, Tohoku Electric was notified that the containment vessel would arrive soon. This put the utility in a potentially embarrassing situation, and it tried to hide the vessel. As one fisherman from Onagawa remarked, "We were on a routine trip and we saw this very large structure on an off-shore island. It was too big to be anything relevant to us. When we discovered it was the containment vessel to house the nuclear reactor, we were shocked and decided to recommence opposition to the plant. We knew we could negotiate much more for selling our fishing rights. What else could Tohoku Electric possibly do with such a structure?"[22] Final transfer negotiations became protracted negotiations, lengthening the licensing process dramatically.

Total lead times to locate initial nuclear (166 months) and fossil-fueled (86 months) packages are longer than total lead times for subsequent nuclear (113 months) and fossil-fueled (51 months) packages. The F-test indicates that statistically lead times for subsequent packages are shorter than those for initial packages. The relative ease in getting consent for subsequent nuclear (28 months) and fossil-fueled (21 months) packages is attributable to shorter public acceptance times compared with initial nuclear (82 months) and fossil-fueled (38 months) packages. In contrast, licensing and

Table 3. Power plant lead times in Japan (months)

Category	Observations	Statistic	Total	Public acceptance	Licensing	Preconstruction	Construction
Nuclear	35	Mean	154	70	22	91	62
		COV	0.4	0.7	0.5	0.6	0.2
Fossil-fueled	55	Mean	75	32	12	43	32
		COV	0.5	1.1	1.1	0.8	0.3
Initial nuclear	27	Mean	166	82	21	103	63
		COV	0.3	0.6	0.5	0.5	0.2
Subsequent nuclear	8	Mean	113	28	25	53	60
		COV	0.2	0.8	0.4	0.6	0.1
Initial fossil-fueled	33	Mean	86	38	15	53	33
		COV	0.5	1.0	1.06	0.8	0.3
Subsequent fossil-fueled	22	Mean	59	21	7	28	31
		COV	0.4	1.0	0.6	0.7	0.3
Fuel		F-ratio[a]	64(6*)	14(6*)	21(3*)	23(5*)	161(2*)
Package		F-ratio	14(3*)	13(5*)	3(6)	16(4*)	1(7*)

Note: [a] A * indicates that result is significant at 99.5.
Source: Survey data.

construction times do not vary much between initial and subsequent packages for any fuel type.

There is consistently more variation in preconstruction lead times than in total lead times. Construction times are relatively more stable. Given that average preconstruction and construction lead times are in general similar, it is the instability in the former that accounts for most of the variation in total lead times. The magnitude of variation in construction times is much smaller and does not greatly affect overall variations.

Substantial variation exists in lead times for both the public acceptance and licensing phases of the preconstruction stage. Variations in public acceptance and licensing times are higher than variations in total lead times. Since, however, licensing times are generally shorter than public acceptance times (particularly for nuclear plants), their influence on total lead-time variation is less significant. Even though bargaining continues into the licensing stage, it does not overly affect the variance in total lead times.

Average lead times increased for the period 1960–1985. Average nuclear lead times were relatively constant during 1965–1975 but increased from 90 months in 1975 to 160 months by 1980. Average fossil-fueled lead times increased from 40 months in 1960 to 100 months in 1980. Increases in average nuclear lead times declined slightly from 1980. Fossil-fueled lead times declined marginally from 1980.

These trends can be explained by distinguishing between lead times for initial packages and subsequent packages. Public acceptance, and not licensing or constructions times, contributes most to total lead-time variations both across space and over time. Before 1980, most of the capacity expansion comprised initial packages, which have longer public acceptance times. From the early 1980s until the present, most of the capacity growth consisted of subsequent packages, which have shorter public acceptance times. Japan has been developing a larger proportion of its incremental capacity (and all for nuclear) on existing sites rather than on green-field ones. Chapter 3 analyzes these contemporary developments in more detail.

The investigation reveals important patterns of lead-time behavior. Nuclear plant lead times are longer than those for fossil-fueled plants. Initial package lead times are longer than for subsequent ones. Variations in bargaining times (and not in licensing or construction times) contribute most to total lead-time variance, both across space and over time. Publicly available data, although not conclusive, point to similar comparative patterns of lead-time behavior in America and Europe.[23] These data provide justification for exploring the nature and role of compensation mechanisms in negotiating siting agreements.

INSTITUTIONALIZED COMPENSATION ARRANGEMENTS

The importance of compensation stems from the relative roles of private versus state actors in siting and the constraints these roles impose on using different policy instruments in managing environmental disputes. Private utilities cannot use eminent domain powers vested in the state to bypass or override local opposition. Consequently, utilities have to rely on compensation instruments to negotiate siting deals with local community interests.

Japan has created very sophisticated compensation schemes for dealing with NIMBY conflicts. MITI has performed only an indirect role in siting, establishing property rights and designing incentive structures that aim to facilitate bargaining between implementing organizations (utilities) and target groups (regional community interests). These schemes provide a set of principles, ensuring that developers pay compensation for the imposition of environmental spillover effects.

As shown in table 4, theoretically, compensation aims to realign the expected cost and benefits of projects and facilitate settlements.[24] This realignment takes place by either increasing benefits or reducing costs (including risks) to aggrieved parties. It can be paid through existing institutions or through other redistributive mechanisms involving special subsidies. Compensation can be offered in both monetary and nonmonetary forms, through both direct and indirect channels.[25] Monetary compensation refers to side payments. Nonmonetary compensation includes risk mitigation, risk substitution, and symbolic compensation, such as agreeing

Table 4. A taxonomy of compensation mechanisms operating in Japan

Mechanism	Compensators	Compensatees	Form of payment
Rights transfer	Utilities	Rights holders	Direct monetary
Subsidies	Utilities	Communities	Direct and indirect
	Taxpayers	Utilities	monetary
Risk mitigation	Utilities	Communities	Indirect nonmonetary
Abandonment	Utilities	Communities	Direct monetary
Development	Consumers	Rights holders	Indirect monetary
		Communities	
Risk reduction	Consumers	Utilities	Indirect nonmonetary
		Rights holders	
		Communities	
Sociopolitical	Utilities	Communities	Direct and indirect
	Governments		nonmonetary
Bribery	Utilities	Individuals	Direct and indirect
		Groups	monetary

Sources: Tsūshō sangyō shō 1963a, 1963b; Shigen enerugii chō 1987; and interviews.

to political compromises.[26] Direct compensation can be paid by developers directly to affected parties. Indirect compensation can be paid indirectly from nonprivate entities to local interests. Compensation might sometimes be offered for not proceeding with project siting.

There are two sets of compensation arrangements operating in Japan: the Compensation Standards and the Three Laws.

Compensation Standards

MITI established the Compensation Standards for Electric Power Development (Compensation Standards) in 1963.[27] During the early 1960s, there was rapid economic growth, and electricity demand increased dramatically. The placement of fossil-fueled and hydroelectric plants, which formed the bulk of electricity generation, encountered increasing local opposition. By 1962 electricity reserve margins were 2.5 percent, far lower than the 8 to 10 percent regarded as optimal by the power industry for insurance coverage against unexpected shortages.[28] Utilities and MITI were very worried about electricity shortages and established the Compensation Standards to facilitate the placement of new plants.

The Compensation Standards provide a set of guidelines and rules that assist in transferring property rights to power companies. Japan has a long tradition, dating back to the Edo Period, of recognizing property rights.[29] Before that, powerful regional individuals controlled fishing grounds, and there were conflicts, sometimes leading to fierce battles, over the use of scarce fishing resources. To manage the common pool resource problem associated with coastal fishing, the Meiji government introduced the Fishing Law of 1911, which legally recognized property rights for different types of fishermen. The Meiji Fishing Laws form the basis of the current system.

Table 5 outlines the main aspects of the Compensation Standards. They cover all rights potentially subject to transfer in the development of power plants. Land rights allow owners to use land for such purposes as agricultural production or investment (capitalizing on it as an asset).[30] Fishing rights permit fishing cooperatives and other individuals to use designated sections of the ocean to catch or cultivate marine life and to use ocean access areas with respect to those activities. These rights include stationary, sectional, and cooperative fishing rights. Utilities must purchase these rights because power plant construction and operation require ocean resources for transportation and cooling purposes. These rights are granted by the prefecture and are absolute. Holders can exclude others from entering areas covered by the rights. Unlike land rights, the transfer of fishing rights requires agreement by two-thirds of the members of the relevant fishing cooperative and prefectural consent.

Table 5. Compensation standards for electric power development

Objectives	To facilitate siting by establishing guidelines for property rights transfer and compensation for losses.		
Property rights	Fishing rights, including rights to stationary, sectional, and cooperative fishing. Land rights, including rights to forest and agricultural land. Other rights, such as customary rights (traditional rights of access).		
Payment	Fishing	$R \div r$	Where R is average net annual earnings (gross annual income from the catch minus average annual expenses); r is a specified interest rate. Calculations to include existing plans and catch reduction.
	Land	$V \times (B \div A)$	In cases where there have been similar transactions in nearby areas: where V is the adjusted price in accordance with the circumstances of transactions in nearby areas; B is the graded value of land to be purchased; and A is the graded value of land purchased in nearby areas.
		$R \div r$	In cases where no similar transactions have taken place in nearby areas: where R is average annual net earnings (see fishing above); and r is a specified interest rate. In cases where similar transactions have taken place, calculations may be also derived from this formula.
	Others		Based on similar cases, taking into account special circumstances.
Time	Compensation shall be based on calculations at the time of contractual settlement.		
Payment method	Payment shall be made to individuals or to organizations when individual losses are difficult to estimate. Payment shall be made in the form of money, but nonmonetary forms may be accommodated.		

Sources: Tsūshō sangyō shō 1963a, 1963b.

There are two methods for calculating payments. The first is based on the expected (discounted) net annual earnings from the property rights.[31] The second is based on "similar situations," whereby other transfer agreements provide a measure of the value of the property rights. The two methods can lead to substantial differences in the estimated value of property rights. Utilities will typically base offers on the first method, whereas property rights holders will normally request an amount based on the second, particularly if that yields a higher value. In many cases, power companies will make payment based on the expected earnings calculation and also pay "cooperation money" to fishing cooperatives to secure transfer. Cooperation money reflects the relative bargaining positions of the parties and what it costs, over and above market valuations, to acquire property rights.[32]

The form of payment depends on the nature of property rights ownership. Payment to land rights owners is generally on an individual basis. Payment to fishing rights holders is on a lump-sum basis to fishing cooperatives, which then decide on the internal distribution. Parties are bound legally to property rights transfer and other agreements signed. They have no recourse to renegotiate even if the subsequent value of the property right changes prior to actual transfer. This is crucial, as U.S. experience illustrates: The possibility of noncompliance to siting agreements can often prevent negotiating process from beginning in the first place.[33]

The Compensation Standards also legally allow utilities to offer other monetary and nonmonetary compensation, both directly and indirectly, to all community interests. Power companies offer subsidies through preferential purchases and employment to increase benefits to communities. They utilize risk mitigation through design and site location changes to reduce project risks. Utilities use information to alter perceptions about expected or real project benefits and costs. Utilities sometimes pay compensation for withdrawing project proposals to ensure that affected communities at least gain something.

Three Laws

MITI established the Three Electric Power Development Laws (Three Laws) in 1974.[34] During the early 1970s, electricity supply could not keep pace with demand and by 1973 reserve margins had fallen to 3.7 percent, which, like in the early 1960s, was far below the optimal level.[35] The emergence of broadly based citizens movements in the late 1960s and early 1970s was crucial in constraining supply growth.[36] The oil crisis caused major concerns about energy security. As one astute observer remarked, "While the oil crisis created energy security worries, Japan would probably have experienced massive electricity shortages in the mid-1970s had it not occurred and forced the economy into recession. We just could not get plants built quickly enough to meet demand requirements."[37] Developments in both domestic and international energy markets combined to move alternative energy expansion to the forefront of energy policy objectives.

By the early 1970s, the management of intra- and intercommunity distributional conflicts had become the most pressing issue in acquiring community consent for energy projects. The Compensation Standards catered mainly to property rights holders. Those who did not have rights felt that those who did were gaining at their expense. Both groups faced environmental risks. Demands for more compensation came from both host communities where negotiations were being conducted and neighboring areas. Power companies were reluctant to pay large amounts of compensation to both host and adjacent communities, both of whom argued that the proposed amounts were not adequate.

Table 6 contains a digest of the Three Laws, a scheme like France adopted and Howard Kunreuther et al. proposed for the United States.[38] The aim of these laws is to redistribute some of the gains from the national community to regional communities. Utilities are taxed on electricity sold, although the national public ultimately bears the burden in the form of higher electricity prices. MITI then channels these funds into a special account under its jurisdiction. Subsidies to host communities are calculated by multiplying a unit subsidy by the plant capacity by a coefficient. In contrast to the Compensation Standards, the use of those payments is restricted to development of community infrastructure and development purposes. Funds are also used to promote the safety of nuclear power.

The amounts payable are greater for nuclear power relative to conventional types of generation. Although unit subsidies are the same for nuclear and fossil-fueled projects, the coefficient is larger for nuclear plants. The

Table 6. Three electric power development laws

Introduction	To tax utilities to provide subsidies to facilitate power plant siting.
Electric Power Development Tax Law	Utilities to be taxed at a rate of 300 yen per 1,000 kilowatt hours of electricity sold.
Electric Power Development Special Accounts Law	Revenue: Electric power development taxes and previous year's balance. Expenditure: Subsidies for provision of social overhead capital based on Regional Development Law (see below) and special subsidies for promoting nuclear safety and effective use of waste water.
Regional Development Law	1. To host communities where power plants are under construction or operating: Facility type　　Unit subsidy × Capacity × Coefficient 　　　　　　　　　(yen per kw)　　(kw) Nuclear　　　　　450　　　　　　　　　　7 Fossil-fueled　　　450　　　　　　　　　　4 Hydro-electric　　120　　　　　　　　　　5 2. To neighboring communities whose boundaries touch the boundary of host communities: each electorate to receive an amount equivalent to the total subsidy to the host community divided by the number of neighboring communities. No subsidies to be provided to neighboring communities for hydroelectric power plants. Period of payment: from commencement of construction to five years after commencement of operation.

Source: Shigen enerugii chō 1987.

larger nuclear coefficient partly reflects the perception that nuclear risks are higher and the increased difficulty in getting local communities to accept those risks. But it also reflects, however, the special emphasis attached to nuclear power in Japanese energy diversification policy.

The Three Laws assume that power plant risks are not confined to administrative boundaries and seek to manage intercommunity distributional concerns. Before their introduction, neighborhood communities tended to receive fewer benefits from siting than hosts, yet both communities faced similar levels of environmental risk. Surrounding municipalities whose boundaries touch those of the host community each receive an amount equivalent to the total subsidy to the host community divided by the number of surrounding municipalities. Payment generally starts with the commencement of construction. This provides an inducement to both host and neighboring communities to negotiate siting agreements and to deter unnecessary delays in licensing. MITI makes payment to the prefecture, which then channels funds to the subprefectural level. Negotiations between utilities and subnational interests determine the internal distribution of funds within communities.

Japan also uses a range of other compensation instruments to manage siting conflicts. These include direct or indirect subsidies to local governments or utilities from either prefectural or national governments; risk substitution, which compensates communities through trading one risk for another; symbolic compensation, such as appointment to regional bodies, and possibly (rumors are rife, and arrests and indictments for bribery and graft in local politics are increasingly frequent) under-the-table deals.

Less comprehensive compensation schemes operate in the United States.[39] Compensation for acquiring land is paid in much the same way as in Japan. Yet there are no institutionalized arrangements for compensating losers for anything other than the transfer of property rights. Developers offer subsidies, contingent compensation, and risk mitigation, but they do so on a case-by-case basis. There are also differences in the ways compensation is paid to localities. Even when developers offer compensation, legally they only have to direct it to host, and not neighboring, communities.[40]

Students of facility siting and public policy have put forward several reasons for the lack of comprehensive compensation schemes in the United States. The first is "logrolling" or the belief that public policy eventually equalizes the overall distribution of costs and benefits.[41] The second is a belief that the potential gains of compensation are lessened by the costs of establishing those mechanisms.[42] The third is that states have opposed federal compensation schemes for siting in order to guard their own autonomy.[43] The fourth is the legal problem of binding parties to compensation agreements.[44] The fifth is that Congress seems to focus more on the source of potential harm, and to leave victims to fend for themselves in state courts with nonuniform and often unfair standards for adjudication.[45]

More extensive compensation arrangements operate in other nations, although they appear to be different in nature from Japan's. In France, public utilities generally reduce electricity tariffs for host communities; they also pay a range of taxes and levies to the host commune as well as to the department as a whole. In Belgium, developers offer funds for industrial infrastructure improvements. In Spain, provinces that export surplus electricity receive payments from the utilities amounting to 5 percent of the value of exported electricity. Canada also uses compensation, although as in the United States, it is ad hoc, varies from one province to another, and is not institutionalized as it is in the French and Spanish cases.[46] Japan is more like Europe than the United States in using compensation to deal with siting conflicts.

CONCLUSIONS

The consistency with which planners in Japan accomplish electricity supply objectives depends heavily on the negotiation of siting deals with community interests. The back-end of siting, or licensing and construction, does not affect the stability of achieving policy objectives. Although bargaining continues into the licensing stage, it is then generally concerned with finer points of negotiations and does not influence variations in total lead times. Construction commences only after all community bargaining processes have been completed, and it is relatively free from political processes.

Compensation is crucial in resolving energy siting conflicts, given that utilities cannot use eminent domain powers and that property rights and other community interests have veto power over siting decisions. Compensation schemes have been established by MITI to deal with spillover effects, but the utilities use them in negotiating directly with property rights holders and other community interests. The various schemes complement one another. The Compensation Standards essentially provide direct monetary compensation to property rights holders and, to a lesser extent, host communities. The Three Laws offers indirect monetary incentives to both host and neighboring communities. Developers also use other forms of compensation, including risk mitigation, subsidies, and symbolic compensation. Taken together, these different compensation mechanisms provide developers with a considerable arsenal of bargaining instruments for addressing the politics of distribution and managing complex siting processes.

The institutional framework that governs siting in Japan is structured to provide strong incentives for utilities to meet electricity demand increases and to facilitate negotiations with property rights holders and other wcommunity interests. Although compensation arrangements help in conflict resolution, they do not necessarily create stability in the time required to negotiate agreements. We need to develop an approach for analyzing the patterns and determinants of resistance to and support for siting and the conditions that facilitate effective use of compensation. Chapter 3 analyzes the importance of the structure of the bargaining environment in explaining observed variations in negotiating times.

C·H·A·P·T·E·R T·H·R·E·E

Structure of the Bargaining Environment

Conflict almost always occurs in siting environmentally hazardous project because projects have different values for different parties. Promoters and developers of energy projects want to achieve market and security goals. Siting imposes an array of both positive and negative spillover effects on community interests. Projects may bring financial and other regional developmental gains. They may also involve health risks, quality-of-life concerns, and other social and political costs. Resistance to projects occurs when communities believe that net spillover effects are negative or when they judge that there is inequality between who benefits and who loses.

Reaching settlements over the siting of energy facilities involves bargaining over an acceptable allocation of the costs and the benefits from those projects. Developers will be required to compensate community interests for losses expected to be incurred from hosting facilities. The value that developers and community interests attach to projects shapes their responses to developing and accepting those facilities and influences bargaining times. These responses will vary depending on the nature of electricity markets and the regional political economy, and the responses are likely to shape the structure of the broad bargaining environment. The bargaining environment may act as a positive or a negative catalyst in resolving conflicts. Negotiating agreements is likely to be less problematic when developers place a lot of importance on expected benefits and when community interests place little emphasis on anticipated costs. Within such environments, a greater economic surplus is likely to be available for developers to compensate community interests for negative spillover effects.

MODEL SPECIFICATION

As table 7 illustrates, multivariate techniques allow a formal testing of relationships between a public acceptance lead-time variable and an array of explanatory variables that characterize the structure of the bargaining environ-

Table 7. Explanatory variables and hypotheses with public acceptance times

Proxy measure	Definition of measurement	Variable	Relationship
Expected electricity shortages in utility spheres	Five-year predicted demand minus expected capacity (installed capacity plus capacity under construction)	ES^{pc}	−
Changing expectations about electricity shortages in regional spheres	Trend growth in regional electricity shortages defined as above	$R.ES^{ng}$	−
Changing expectations about regional social and economic opportunities	Trend growth in prefecture per-capita income	R.IN	+
Autonomous ability of local government to supply public goods	Local Financial Index (ratio of tax revenues from local sources to expenditure on public goods)	LFI	+
Importance of primary sector in prefectural product	Primary prefectural per-capita product as a proportion of total per-capita product	PC	+
Importance of primary sector in prefectural employment	Persons employed in prefectural primary industry as a proportion of total persons employed	PE	+
Leftist political party representation in prefectural assembly	JSP and JCP[a] seats as a proportion of total prefectural assembly seats	LI	+
Attitudes toward environmental preservation	Time (SA) = 0 if 63 < SA < 69; = 1 if 70 < SA < 79	SA	+
Risk associated with different technologies	Fuel (F) = 0 if nuclear; = 1 if fossil-fueled	F	−
Familiarity with risk	Package (PN) = 0 if initial; = 1 if subsequent	PN	−

Note: [a] Japan Socialist Party and Japan Communist Party.

ment. Several considerations guided the specification and selection of these variables. The first was case study analysis, which identified the major players involved in negotiations and developed propositions about their likely responses to energy projects that could be quantitatively tested. For example, expected electricity shortages (ES) provided one measure of the urgency to develop capacity. When other things are equal, greater projected shortages would probably be associated with quicker approvals because developers would be more willing to provide adequate compensation to resolve conflicts. The second consideration was the compilation of a data set large and homogeneous enough to provide a reliable basis for estimating relationships. This data set included information about electricity shortages and social and economic conditions at different community levels. The third consideration was data availability. Useful information, such as amounts of compensation, was not available to the extent that it could be used statistically.[1]

This analysis covers forty-eight power plants developed in Japan from 1960 to 1979, the period in which most energy plants were constructed. Electricity market data were provided by Tokyo Electric. Community data were gathered from published statistical sources and from the Ministry of International Trade and Industry (MITI).[2] The sparseness of these data for some cases during the 1960s prevented examination of all ninety cases reported in chapter 2. Ten abandoned nuclear cases were used later in the analysis to consider the utility of the models. The sample provides extensive coverage of siting experience in postwar Japan.

The investigation treated costs and benefits in particular ways. First, it distinguished costs and benefits perceived by potential or actual players from those perceived by an analyst, even a well-meaning analyst; it is what players think projects will do for them, and at what cost, that will determine responses. Second, the investigation did not assume that the factors influencing the costs and the benefits are simply income-related; they are also related to other economic (financial health and economic structure) and noneconomic (ideology and attitudes) variables. Third, it did not presume that costs and benefits can be measured in similar ways to all parties; different social, economic, and political conditions may influence perceptions of costs and benefits. Fourth, it did not treat costs and benefits as static; perceptions or expectations change during conflict resolution.

Several model variants were generated. The *predictive model* considered how much of the variation in approval times was explained given conditions at the beginning of bargaining processes. It took the value of variables at the time of a siting decision, the average value for a three-year period before that decision, and trend growth for that period. The *explanatory model* analyzed the degree of variance accounted for by changing economic and political conditions during the negotiating period. Although the specification of this model was the same as the predictive model, the average and trend growth values of the variables differed in that they were measured during the entire approval process. The *evaluative model* combined the predictive and explanatory models and sought to explain variations, taking into account conditions at the beginning of bargaining and those that emerged during resolution. I report the results of only the predictive and evaluative models here.[3]

The analysis ideally should treat projects with different technical characteristics in separate regressions. A relatively small number of observations for initial and subsequent nuclear packages prevented separate treatment. I handled this concern by considering different plants, with the use of dummy variables indicating fuel category and package number, and interacting those dummy variables with explanatory variables. For example, I assigned a value of zero or one to the dummy fuel category variable, multiplied that by a relevant explanatory variable (per-capita income), and then ran the regression. The estimated coefficient showed the combined effects of income and fuel category on negotiating times.

A stepwise regression method chose the final variables. For the predictive and explanatory models, it entered all of the explanatory variables and successively eliminated those that were not statistically significant at the 0.005 level. It then entered significant variables interactively with the dummy variables and eliminated meaningless variables in the same way. The evaluative model combined the predictive and explanatory models and then eradicated trivial variables. Variables measuring prefectural energy self-sufficiency, the number of persons leaving local communities, the timing of Environmental Impact Assessment legislation, and crisis periods defined by Kent Calder were not statistically significant.[4]

POLIMETRIC RESULTS

Tables 8 and 9 illustrate the results of the inquiry. The F-ratio, a test of the joint statistical significance of all of the coefficients, shows that the null hypotheses of no relationship between public acceptance times and the explanatory variables can be rejected for both models. There appears to be predictability in the pattern of varying responses by developers and community interests to energy projects and the impact of those responses on bargaining times. These responses are conditioned by perceived spillover effects of projects and shape the negotiating environment. The structure of the bargaining environment is related to the market need for projects and the structure of the regional political economy.

Another major conclusion was the higher R^2 for the evaluative (74 per-

Table 8. Predicting public acceptance times

Variable[a]	Proxy measure of variable	Coefficient[b]	T-ratio
R.ES^{ng}	Growth in expected regional electricity shortages	−0.01	3.17
IN*F	Combined effect of IN and fuel category (F): F = 0 if nuclear. F = 1 if fossil-fueled.	2.44	7.10
LI	Ratio of JSP and JCP[c] assembly seats to total assembly seats	8.13	12.37
LI*F	Combined effect LI and F	−7.14	5.32
PC*F	Combined effect of PC and F	−8.96	7.00
LFI	Local Financial Index	−0.44	4.04
LFI*PN	Combined effect of LFI and package number (PN): PN = 0 if initial. PN = 1 if subsequent.	−1.00	5.39
	Constant	6.43	
	R-square	0.55	
	Standard error	0.75	
	F-ratio	6.00	
	Observations	48	

Notes: [a] Variables prefixed by R indicate trend growth in those variables three years prior to siting decisions.
[b] The coefficients are measured in log of months per unit of the variables.
[c] Japan Socialist Party and Japan Communist Party.

cent) compared to the predictive model (55 percent). The assessment of approval times can be enhanced significantly by accounting for conditions that emerge during siting conflicts. Predictions at the beginning of the settlement process can be made by considering regional electricity shortages, per-capita income, Leftist party seats, primary per-capita product, and the Local Financial Index (LFI). The influences of Leftist party seats, primary sector production opportunities, and the local financial situation continue to shape the structure of the bargaining environment. Conditions at the beginning of negotiations appear to be swamped by changed utility shortages, primary sector employment opportunities, and social attitudes during bargaining processes.

Utilities are legally responsible for developing adequate capacity to meet expected demand increases in their electricity spheres. As expected, the value that developers place on plants influences their willingness to resolve siting conflicts.[5] Settlement times are likely to be shorter when utilities anticipate large shortages ($-ES^{pc}$). To avoid costly electricity shortages, utilities are likely to place more emphasis on additional capacity and less priority on capital and other costs. Under these conditions, developers are likely to be more enthusiastic about offering adequate compensation to community interests.

Table 9. Evaluating public acceptance times

Variable[a]	Proxy measure of variable	Coefficient[b]	T-ratio
ES^{pc}	Expected utility electricity shortages	−0.26	11.10
$R.ES^{ng}$	Growth in expected regional electricity shortages	−0.02	14.08
$R.ES^{ng}*PN$	Combined effect of $R.ES^{ng}$ and PN: PN = 0 if initial. PN = 1 if subsequent.	0.03	14.45
R.IN	Growth in prefectural per-capita income	−1.23	31.27
R.IN*F	Combined effect of R.IN and fuel category (F): F = 0 if nuclear. F = 1 if fossil-fueled.	0.91	13.40
$P(LFI*PN)^b$	Combined effect of LFI and PN	−0.51	5.02
P(PC*F)	Combined effect of PC and F	−8.79	10.80
PE*PN	Combined effect of PE and PN	4.80	6.90
P(LI*F)	Combined effect of LI and F		
SA	SA = 0 if 63 < SA < 69.	−3.61	9.02
	SA = 1 if 70 < SA < 79.	1.22	24.00
SA*PN	Combined effect of SA and PN	−0.01	6.69
	Constant	4.98	
	R-square	0.74	
	Standard error	0.56	
	F-ratio	10.55	
	Observations	48	

Notes: [a] Variables prefixed by P indicate variables from the predictive model that consider conditions at the time of utility siting decisions.
[b] The coefficients are measured in log of months per unit of the variables.

In contrast to utilities, MITI is concerned with managing broader regional electricity spheres. Negotiating times are likely to be shorter when shortages in these spheres are developing ($-R.ES^{ng}$). An interesting result is that MITI appears to place more emphasis on initial, rather than on subsequent, packages in periods of high expected electricity shortages ($-R.ES^{ng}*PN$). Average public acceptance times for initial packages (126 months) are significantly longer than those for subsequent ones (86 months). In periods of excess demand, MITI appears to be more determined to see initial packages developed quickly because subsequent packages can then be developed relatively more easily. As we will see later, this emphasis on initial sitings also has a role in community addictions to subsequent packages. MITI may seek to expand long-term site availability in increasingly tight electricity markets. It appears more willing to support projects for Electric Power Development Coordination Council (EPDCC) approval when markets are tight, even if compensation agreements are not finalized.[6]

Project siting requires community consent. Even if utilities and MITI urgently want to develop energy projects, the magnitude of spillover effects will influence community responses and the speed at which bargains can be reached. Communities are more likely to accept a compensation proposal in return for their consent when adverse impacts are relatively small.

Prefectural communities appear to accept power plants more readily when per-capita incomes are rising relatively rapidly ($-$R.IN). This finding does not support the hypothesis that the demand for environmental quality is likely to be low and, hence, there is likely to be less resistance to unwanted projects when incomes are low.[7] A boom-economy mentality usually emerges under conditions of rapid income expansion, and there may be a lag in anticipating negative consequences of energy projects. Communities probably want social and economic benefits associated with economic booms to continue and may be less resistant to energy plants. As a result, they may require less compensation in exchange for projects.[8] It may not be relative levels of incomes, but rather expectations about incomes, that influence the demand for environmental quality.[9]

The importance attached to environmental quality appears to be a stronger factor in lengthening negotiations for conventional, compared with nuclear, projects ($-$R.IN*F). In contrast to nuclear plants, fossil-fueled projects are usually located close to major consumption areas where there are relatively high incomes and pollution levels. Communities are likely to place less weight on income benefits associated with fossil-fueled plants and more weight on expected air pollution costs. Emissions from fossil-fueled plants are more visible than those from nuclear plants.[10] Nuclear risk may be perceived as less harmful by communities experiencing a boom in economic activity.

Local government interest in energy siting is related to the financial state of their communities as accepting power plants yields significant financial benefits, such as tax revenues. The relationship between the LFI and public acceptance times does not support the hypothesis that settlement times are longer when the financial state of local governments is relatively healthy. This variable does not capture important effects that initial project development has on the financial state of local economies. Project siting strengthens the local tax base for developing social overhead capital. Financial benefits tend to decline after construction, and maintenance costs on social capital impose a severe drain on financial resources.

The effect of the financial health of local governments is stronger in shortening settlement times for subsequent packages relative to initial ones ($-$LFI*PN). Settlement times are likely to be shorter for subsequent packages because of the financial difficulties experienced after constructing the initial packages when taxes and other revenues generated by that construction decline quite dramatically. Local governments are likely to be more eager to fulfill community expectations for continued growth and to be less worried about further environmental consequences.

This financial bias is one crucial factor explaining the tendency to concentrate power plants in specific locations in Japan. As one student of Japanese environmental politics pointed out, "Energy facilities may be one of the few types of large-scale public works projects where there can be good reason, from the perspective of both the developer and local community, to build more of the same after the first one is built."[11] Contrast this power plant concentration with airports. Niigata may need one international airport and a *Shinkansen* line, but it certainly does not need seven, the number of nuclear plants Tokyo Electric has built at Kashiwazaki-Kariwa, making it the world's largest nuclear site.

Development of facilities almost always requires the transfer of property rights possessed by primary sector groups. Rural opposition appears to be a stronger factor in lengthening negotiating times for nuclear, compared to conventional, plants when the primary sector is more important in generating regional product ($-P.PC^*F$). The rural sector appears to be more concerned about nuclear risk. It generally perceives devastating effects on primary industry from any nuclear accidents and discharge of radioactive substances. Even though emissions from fossil-fueled projects can also have severe effects, they seem less likely to worry primary sector interests.

A second aspect of rural responses is that primary sector employment opportunities seem to be more important in delaying bargaining processes for subsequent packages relative to initial ones ($-PE^*PN$). Rural interests usually oppose initial packages, as they fear the risks involved. They appear to become more anxious about the possible risks of more power plants on factors of production, such as land and ocean resources. The responses of primary interests to accepting nuclear plants appear to be different from those of the community in general under different economic conditions.

Ideology can affect siting processes. Longer bargaining times are associated with stronger Leftist party representation in prefectural assemblies ($+LI$).[12] Leftist political parties tend to place more emphasis on protecting the environment and usually attempt to sensitize the public to the risks of power plants.[13] The importance of ideological orientation appears to be stronger in lengthening bargaining processes for nuclear, compared to fossil-fueled, projects ($-LI^*F$). Japan's Leftist parties obtain electoral support from the coal industry in return for protectionist policies.[14] They are likely to be less concerned about the environmental costs of fossil-fueled projects.

Changing community attitudes in the late 1960s toward more emphasis on environmental preservation is associated with longer negotiating times ($-SA$). The effect of this transformation of attitudes appears to have been less important in delaying approval times for subsequent packages relative to initial ones ($-SA*PN$). Although accidents and pollution problems continued to occur after the early 1970s, community familiarity with the risks of having power stations operating appears to facilitate negotiating processes for subsequent packages.

An unexpected finding is that the regression results do not generate any statistically significant coefficients for fuel (F) category and package number (PN) as variables by themselves. This is puzzling, particularly in light of the results of the previous chapter, which found that nuclear plant settlement times are longer than fossil-fueled plant ones, and initial package approval times are longer than those of subsequent packages. The key reason for this anomaly is that dummy variables are rather crude in a statistical sense and that variations in other social, political, and economic variables are more useful in explaining bargaining times.[15]

In contrast, both F and PN are meaningful in combination with other explanatory variables, which suggests, at least tentatively, that the evidence supports neither the proposition that nuclear plants will always take longer to site than conventional projects because of the inherent nature of nuclear risk nor the belief that initial power plants will always take longer to develop than subsequent power plants because of the unfamiliarity with risky environments. The significance of F and PN combinations supports the contention that the importance attached to the risk of projects with different characteristics is likely to depend on social, economic, and political conditions.[16] In evaluating community responses to power plants, there is a need to go beyond technological features and to examine the structure of the political economy.

The structure of the bargaining environment is an important determinant of the time taken to resolve siting disputes. These structures are likely to be more conducive to quicker settlements when developers have strong market needs for projects and when communities are not overly concerned with any adverse spillover effects. In summary, conflict resolution times will be *shorter* when

utilities expect large shortages in their electricity spheres;
MITI foresees large electricity shortfalls in broader regional electricity spheres, particularly for initial packages;
prefectural incomes are expanding rapidly, particularly for nuclear plants;
the local financial base is weak, particularly for subsequent packages;
prefectural Leftist party representation is relatively small, particularly for conventional projects;
economic opportunities in the prefectural primary sector are relatively scarce, particularly for fossil-fueled projects and initial packages; and

social attitudes place a relatively strong emphasis on environmental preservation, particularly for subsequent packages.

FURTHER QUANTITATIVE INSIGHTS

Although the models are statistically significant and explain a large proportion of the variation in bargaining times (particularly the evaluative model), a closer examination of residuals provides further insights into energy facility siting in Japan.

Predictive and Explanatory Power

As shown in table 10, the investigation compares *arbitrary public acceptance times* (arbitrary PAT) for ten nuclear power plants, excluded from the earlier analysis, with times estimated by the models. It defines arbitrary PAT as the time from a power company's siting decision to 1979, the cut-off point for the regression analysis. The statistical analysis of unapproved plants is difficult. Excluding unapproved plants creates a sampling bias because certain disputes are not considered. Including these sites distorts the results because settlement times are actually infinite. One way to incorporate those cases is to exclude them from the models and then to use the estimated coefficients to generate predicted times for them. This method permits a rigorous test of the predictive and explanatory power of models, particularly as it uses the most problematic observations in the entire data set—the abandoned developmental proposals.

The analysis considers the magnitude and sign of the *PAT error*, the difference between arbitrary (as opposed to actual) settlement times and times predicted by the models. The models are useful for predicting and evaluating if results yield large and negative PAT errors (considerable overestimation of arbitrary PAT), as settlements would have been protracted well beyond 1979. In contrast large positive PAT errors (considerable underestimation of arbitrary PAT) imply that settlements would have been reached before 1979. As we know, that these projects were abandoned, positive PAT errors suggest that the models are not taking into account adequately special characteristics of these disputes that make them significantly different from observations contained in the regression sample.

The predictive model underestimated considerably arbitrary settlement times. Most of the observed errors, such as those for Ashihama (165 months), Kumano (129 months), and Noto and Hidaka (106 months), were

Table 10. Predictive and explanatory power of the models

Plant	Utility	Period	Arbitrary PAT	Predictive model[a]		Evaluative model[b]	
				Predicted	PAT error	Predicted	PAT error
Ashihama	Chubu	1963–1979	179	14	165	87	92
Shimokita	Tohoku	1963–1979	159	61	98	100	59
Namie	Tohoku	1967–1979	140	81	59	99	41
Noto	Hokuriku	1967–1979	140	60	106	107	33
Kumano	Chubu	1967–1979	140	11	129	126	14
Hidaka	Kansai	1967–1979	141	35	106	96	45
Koza	Kansai	1967–1979	140	35	105	96	44
Nachi-Katsuura	Chubu	1969–1979	110	25	85	103	7
Suzu	Hokuriku	1975–1979	45	13	32	145	−102
Hikigawa	Kansai	1976–1979	30	40	−10	154	−124

Notes: [a] The predictive model examines the explained variation given conditions at the beginning of bargaining.
[b] The evaluative model combines the predictive and explanatory models. It explores the explained variation given conditions at the beginning of negotiations and conditions that emerge during the course of settlement. The explanatory model analyzes the explained variation given changing conditions during the approval process.

large and positive. The evaluative model also underestimated settlement times in most cases but to a significantly lesser extent. The largest positive PAT error is Ashihama (92 months), significantly less than the one derived from the predictive model. The evaluative model does, however, yield two cases with very large and negative PAT errors: Suzu (-102 months) and Hikigawa (-124 months).

An important caveat of the models was the tendency to underestimate approval times for projects facing negotiating difficulties in the earlier period of the analysis. For instance, PAT errors for Ashihama (1963–1979) were large and positive, whereas those for Hikigawa (1976–1979) were large and negative. When utilities confronted strong resistance, making bargaining costs high, they often changed site priority rankings. They tended to give priority to other, less difficult sites, and either delayed attempts at negotiating at original sites or reassigned them to candidate site status. This evidence, however supports the earlier argument: although there is some predictability in settlement times, conditions that emerge during the settlement process are crucial in determining actual negotiating times.

All of these abandoned disputes were more problematic than those in the regression sample. In general, they occurred in areas in which there were extremely rich fishing (Ashihama, Suzu, Noto, Koza, Hikigawa, and Hidaka) or farming cooperatives (Shimokita, Namie, Nachi-Katsuura, and Kumano). Intense opposition conditioned bargaining environments in which utilities tried to resolve conflicts. Even though some larger utilities (Kansai and Chubu Electric) faced large electricity shortages, other smaller ones (Tohoku, Hokuriku, and Shikoku Electric) expected only small shortages. Utilities, independent of their electricity market conditions, were unable to offer adequate compensation to negotiate settlements.

Variations in Residuals

The coefficient of variation of the residuals (a measure of dispersion around average residuals) for both nuclear and conventional plants is significantly lower for subsequent packages (3.19) relative to initial ones (9.04). The models explain bargaining times consistently better for subsequent packages. This variation suggests more stability in terms of the effects of bargaining environments on settlement times for developing subsequent packages relative to initial ones.

Utilities find it more difficult to break community resistance the first time but they find it easier later, and they are aware of this.[17] As electricity demand continued to increase during the 1980s and 1990s, utilities gave priority to building nuclear projects at subsequent sites that had lower bargaining and compensation costs. As a result, they often delayed negotiating settlements with communities at higher cost, green-field sites.

The impact of large projects on small, local communities is enormous. Growth in demand for goods and services is rapid during construction. Incomes, financial reserves, employment, and other social and economic opportunities expand markedly. When construction is finished, these benefits diminish abruptly as the completed plant in itself does little to provide a base for sustained regional development.[18] This economic boom-and-bust phenomenon has become more acute since the mid-1970s, when Japan entered a slower growth era. Leftist political party and primary sector interests found it more difficult to oppose subsequent projects. Community interests generally became more familiar with the risks of having subsequent projects, and they attached more weight to the benefits of subsequent projects (to prevent regional economic recession) and less to the environmental costs.

Utilities generally purchase adequate property rights to develop several plants rather than negotiate on a plant-by-plant basis. As the Compensation Standards allow for no legal recourse for renegotiating once initial agreements are reached, approval processes for subsequent projects do not normally entail drawn-out negotiations over property rights transfers. Given the boom-and-bust phenomenon, communities were probably more willing to accept less community compensation for consenting to continued developments. Yet, from 1974, with the establishment of the Three Laws, communities were provided the same benefits for subsequent projects as they were for initial ones. State institutions provided further incentives for communities to accept subsequent projects.

One power company official summed up nicely community addictions to subsequent plants after the first ones are built:

> During the last decade or so, many local governments approached utilities requesting us to build power plants in their localities. Rather than asking local communities for approval, as we did in the 1970s, those communities have been asking us to build facilities. Some have even put a lot of political pressure on us through prefectural and national politicians. This has created a problem for the industry. It is often difficult to justify projects because of slow economic growth and recessed electricity demand. But we cannot say to them [local governments] that we do not want to build them. We have always argued the necessity for more power plants. If we decline, they, and the anti-nuclear movement, may use that to oppose projects in the future should we need to build more plants. One of our biggest problems has not been how to develop power plants, but rather how *not* to develop too many power plants. It is hard to actively promote power plants and simultaneously appear to oppose them.[19]

These structural biases arising from the economic boom-and-bust phenomenon and utility property rights acquisition strategies have created a "reverse NIMBY syndrome" in Japan during the 1980s and 1990s. This syndrome is the most important factor explaining why no green-field nuclear sites have been opened since the mid-1980s and why all developments since then have occurred at sites where plants were already operating. The structure of the bargaining environment after initial conflicts are resolved provides utilities and communities with strong incentives to continue developing more power plants. Almost all of the nuclear plants that have been built during the last fifteen years have been developed on sites initially opened in the 1960s and 1970s.

The siting of initial nuclear projects has become much more difficult since the mid-1980s. One measure of the magnitude of the difficulties facing the nuclear power promoters can be highlighted by considering an extremely optimistic scenario of all previously abandoned nuclear plants getting approval in 1998. Should this happen, the average bargaining time for initial nuclear projects would be thirty years, ranging from twenty-two years (Hikigawa) to thirty-five years (Shimokita). This bargaining time represents more than a twofold increase from the early 1980s, when the average was twelve years. The recent 1997 local rejection of the initial Maki nuclear project, which Tohoku Electric had been promoting for about twenty years, reinforces this conclusion. Japan's nuclear developers now face public acceptance times of at least twenty to thirty years at initial sites.

CASE STUDIES

Explaining variations in settlement times need and should not rely only on statistical results, even very good ones. A combination of polimetric and case study approaches generates a more comprehensive and robust explanation of siting conflicts than would be possible simply by using one approach or the other. This combination approach would be valid even if statistically significant models explained all of the observed variations in bargaining times.

Table 11 summarizes the cases identified for closer examination. Case selection was based mainly on the size and stability of the residuals and, as a result, only initial packages are included. The final choice of these cases was based on several other criteria, ensuring the sample was relatively representative: cases covering different time periods and regions across Japan; one case with a small residual (Hamaoka) to allow comparison of cases with large ones; one case over a fossil-fueled plant (Matsushima) to allow comparison between fossil-fueled projects in relatively urban areas with

Table 11. Key descriptive features of case studies

Characteristics	Power plant				
	Ashihama	Hamaoka	Matsushima		Tomari
Company	Chubu	Chubu	EPDC		Hokkaido
Plant	Initial nuclear	Initial nuclear	Initial coal		Initial nuclear
Site					
Prefecture	Mie	Shizuoka	Nagasaki		Hokkaido
Local	Initially on Nanto and Kisei (border), then in Kisei	Hamaoka	Matsushima island in Ooseto		Initially in Kyowa and Tomari, then in Tomari
Period (PAT)	1963–(shelved in 1967)	1967–1969 (23 months)	1973–1976 (44 months)		1969–1982 (156 months)
Key features	Opposition from fishing interests, local public, and conservative politicians	Opposition from fishing groups, local public, and landowners	Opposition from private utility, MOF, fishing groups, and landowners		Opposition from fishing groups, the coal industry, and other utilities
	Opposition gained access to power through recall movements	Leaders and power brokers split opposition alliance	Promoters dealt with utility and MOF but problems with other interests		Utility confronted high compensation and management costs
	Deal not reached; high compensation offered	Deal struck very quickly; little compensation	Deal took moderate time; very high compensation		Settlement took long time; very high compensation

Utility projection[a]	25 (154)	6 (17)	24 (20)	40 (116)
Predictive model	14 (165)	24 (−1)	68 (−24)	110 (46)
Evaluative model	87 (92)	15 (8)	68 (−24)	131 (25)
PAT error: Evaluative model	Much longer than model infers	Slightly longer than model infers	Moderately shorter than model infers	Moderately longer than model infers
Compensation[b]	7,600 (offered)	1,860	13,640	13,800
Compensation standards				
Fishing (per-capita)	900 (0.6)	600 (0.2)	1,640 (2.3)	3,050 (2.1)
Land	Little	Moderate	Moderate	Little
Community[c]	0	1,260	6100	4,210
Risk mitigation	Major site change	Minor design change	Minor design change	Major site change
Abandonment	Not offered	Not applicable	Not applicable	400
Three Laws	Not applicable	Not applicable	4,250	6,140
Other				
Prefectural subsidies	6,700 (offered)	Not offered	Not offered	Not offered
Sociopolitical	To local interests	To regional interests	To local interests	To regional interests
Bribery	Reportedly to politicians	Reportedly to individuals	Not reported	Not reported

Notes: [a] Brackets indicate errors. The Ashihama error is calculated using 1979 as cut-off date. As of 1997, it still had not obtained EPDCC approval.
[b] Total quantifiable (in million yen). These estimates exclude compensation for land rights.
[c] Includes "cooperation money" mainly directed to fishing cooperatives. At Matsushima, it was mainly channeled to Kyushu Electric.

nuclear plants in relatively rural areas; at least one case where predictions overestimate (Matsushima) or underestimate (Ashihama and Tomari) bargaining times to permit comparisons of factors influencing the sign of residual errors; one case representing abandoned projects (Ashihama) to allow comparisons with cases where agreements were eventually made, no matter how short (Hamaoka) or long (Tomari) a period they required; and two cases involving the same utility (Hamaoka) to permit a comparison of strategies and lessons learned over time.

Chapter 4 focuses on the Ashihama dispute, where opposing economic and political interest groups were able to gain electoral access to local government and forced abandonment of a project.

Chubu Electric attempted to locate the Ashihama nuclear plant on the border of Nanto and Kisei towns in 1963. It faced strong resistance from Nanto fishing cooperatives, which were experiencing an economic boom, and could not compensate them adequately. The opposition gained electoral access to local power. Subsequent prefectural and national government pressure only intensified resistance. The utility then changed the location of the plant to a site solely within Kisei town. It did not structure its compensation policy to take into account intra-party competition in the Kisei Liberal Democratic Party (LDP). The opposing faction mounted a successful recall movement. In 1967 Ashihama became the first nuclear project abandoned in Japan.

Chapter 5 explores the Hamaoka conflict, where promoters broke apart an alliance of convenience between fishing and ideological interests.

Chubu Electric, after the Ashihama debacle, located its first nuclear plant in Hamaoka in 1969. Regional fishing cooperatives formed a coalition with a Leftist anti-nuclear movement and attempted to oppose the project. Backed by the utility, the regional political leadership quickly split the alliance. This prevented the Leftist movement from overly hampering negotiations over compensation and property rights transfer arrangements. They then removed the leader of the fishing cooperative, who was opposing the project, and installed a new leader who was eager to negotiate a quick settlement.

Chapter 6 details the Matsushima case, where a public utility negotiated with a private utility and other bureaucratic interests before dealing effectively with the local community.

The Electric Power Development Company (EPDC) won consent for the Matsushima coal-fired plant on a small island called Matsushima, off Ooseto town, in 1977. Unlike many initial siting conflicts, there was no organized resistance at the local level. The EPDC, however, faced opposition from Kyushu Electric and the Ministry of Finance (MOF) because of the project's high expected costs. In response to the oil crisis, national policy started to stress energy security and regional electricity objectives. The increased importance of the project allowed the national government and other regional utilities to provide adequate subsidies and compensation to private and community interests, thus striking a settlement.

Chapter 7 investigates the Tomari conflict, where increasing project costs forced a utility to significantly delay negotiations with local community interests.

Hokkaido Electric received approval to site the Tomari nuclear plant more than a decade after it approached the local community in 1969. Economic and ideological opposition forced the utility to delay the project at various stages because of increasing costs of developing nuclear power. Unlike Matsushima, even the oil crisis did not alleviate these high compensation costs. In response, the utility initially delayed the project but, due to a terrorist threat on the plant, it changed the location of the plant and finally reached a settlement.

The first objective of the case studies is to explore how polimetric estimates of settlement times fare alongside utility ones. I derived utility estimates from local newspapers and confirmed them in interviews with company personnel. These forecasts do not necessarily mean what they say and are not necessarily accurate. They sometimes include strategic elements. Utilities may publicize forecasts to exert pressure on local interests, to determine potential patterns of resistance, or to justify obtaining subsidies from the state.

Bearing this in mind, as shown in table 11, company forecasts underestimate approval times, particularly for the Ashihama and Tomari projects. The predictive model provides estimates that were equally as accurate or better than utility assessments. These comparisons suggest that the models, subject to our earlier qualifications, provide better estimates of approval times than Japanese utilities themselves. Power companies would fare better if they complemented least-cost site selection approaches with political economy ones to identify sites where negotiating agreements is more likely.

The case studies show that despite strategic elements in utilities' forecasts, utilities usually do not gauge the potential intensity of, or their ability to manage, resistance. At Ashihama, the utility planned for a two-year settlement period but ultimately abandoned the project. At Matsushima, where the utility expected opposition from another private utility, its forecast was still off by about two years. Even though Japan's approach to siting assumes that compensation is necessary, the mere existence of redistributive mechanisms cannot substitute for careful planning and management of negotiations.

A second objective is to examine the errors of the models. Table 11 shows that the direction and magnitude of the errors vary widely. As expected, other factors, such as the distribution of costs, the structure of bargaining power, bargaining strategies, changing expectations, and uncertainty about outcomes, also condition bargaining in complex and crucial ways. For instance, the regression assumes that utility strategies are similar in all conflicts. Utilities may, however, design different strategies, and the skill with which they execute those strategies will also influence dispute resolution. The existence of these factors, which are not easily quantified, increases dramatically the variance of the data set.

The case studies seek to explore the relationships among these factors and the times required to obtain community agreement for projects. For example, the distribution of costs across different fishing cooperatives was spread more evenly across those cooperatives at Ashihama than they were at Hamaoka. Even though fishing groups put up intense opposition initially at both sites, they were clearly more effective in sustaining collective action at the former. Through case comparisons, it is possible to assess the nature and strength of the relationships among these variables, that determine the errors of the models, and to develop broader generalizations about their influences on settlement times.

A third aim of the case studies is to provide a better understanding of the relationship between process and outcome. Regression analysis only provides information (although very important) about input–output relationships. For example, at Hamaoka it suggests that the approval time was very short because promoters placed a high value on the benefits of the project and the community was not overly concerned about environmental costs. This combination enabled the power company to provide adequate compensation to effect a settlement. Although this perspective is very important, it does not reveal much about the political processes involved in that compensation. We can further our understanding of siting by analyzing how variables, such as the structure of the bargaining environment, expectations, and uncertainty, interact with each other, and the ways in which those interactions are translated through political processes into negotiated siting outcomes.

The final objective is to explore the compensation issue. Compensation is used extensively in siting disputes in Japan. It is essential to shed light on relative use of different forms and how they fit together in the resolution of siting disputes. Compensation requirements for initial projects have increased, and the nature of institutionalized compensation mechanisms has changed. Examining who was compensated, by how much, through what mechanisms, and in what forms is crucial for assessing who wins and who loses and who has real bargaining power in Japan. We can also understand the conditions under which compensation is likely to be effective in managing facility siting and get some idea of what it will likely cost utilities to resolve siting conflicts in the future.

CONCLUSIONS

Unfortunately, NIMBY conflicts do not always fit comfortably into a narrow spectrum of difficult and protracted cases. If communities regard siting as a "bad bargain," and if they use "collective action" or "protest as a political resource," then why do we observe such marked diversity in negotiating times? Communities have delayed significantly some projects and have killed others, but they have delayed others very little and in some cases have actually encouraged utilities to build projects in their backyards.

Polimetric analysis provides a first and very important attempt to explain these variations. A crucial part of the answer lies in the nature and structure of bargaining environments. Project siting always imposes costs and benefits on developer and community interests, but the magnitude of those impacts varies, thereby influencing the speed at which developers receive community consent. Particular bargaining environments will be more conducive to mitigating protest and collective action than others. Negotiating environments, characterized by the large perceived economic surpluses that can be used to compensate communities for spillover effects, are likely to result in quicker conflict resolution.

Not all delays can be simply ascribed to community resistance. Similar levels of community resistance can lead to very different bargaining times. Long delays may be encountered when projects are expected to impose large negative spillover effects on communities, if developers have little market need for projects, or if they do not provide adequate compensation. At the other extreme, agreements may be resolved quickly, even when communities expect significant negative impacts, if there is a strong requirement for additional projects and developers provide sufficient compensation. Grounding our analysis in bargaining theory forces us to investigate explicitly both developer and community responses and their impacts on settlement times.

The quantitative models are statistically significant and account for a large part of the variance in bargaining times. They offer considerable predictive and explanatory power even with a relatively small number of variables. Knowing the broad structure of the bargaining environment provides us with predictions that are generally better than the utilities' own forecasts. Understanding structural biases in those environments associated with the economic impacts of initial projects on local communities and property rights acquisition strategies explains why subsequent projects are consistently easier to site. These explanations are crucial in understanding contemporary siting developments and why Japan has been able to site only subsequent nuclear projects since the mid-1980s.

Even though the quantitative models generate powerful conclusions, they need to be supplemented and integrated with qualitative approaches. The remainder of this book details the richness and complexity of several energy siting controversies in Japan. By doing so, it provides further analytical insights into bargaining and compensation processes involved in managing NIMBY disputes.

C·H·A·P·T·E·R F·O·U·R

Gaining Access to Political Power

Negotiating siting outcomes requires consent of different community groups. Those who expect to be adversely affected by projects will oppose them to enhance their positions. The extent to which those groups mobilize, gain access to political power, and alter the structure of the bargaining environment has a decisive impact on bargaining outcomes. Bargaining is more likely to end up in gridlock, where opposing interests control or gain control of local decision-making centers, unless these groups can be removed from power.

Ashihama is interesting because it provides an extreme case of how local communities can reject unwanted projects. Although nuclear capacity grew rapidly since the 1960s, more than half of the planned commercial nuclear projects during this decade were shelved because of local opposition. Ashihama was the first.[1] It is a classic NIMBY case. Disputes where projects are ultimately abandoned are fundamentally different from those where agreements are made no matter how long it takes to resolve them. Such cases provide a benchmark against which to consider cases where compensation mechanisms are more effective in negotiating processes. They also illustrate the difficulties utilities have faced in opening new sites during the latter half of the 1980s and during the 1990s.

What makes Ashihama puzzling in contrast with other cases examined in this book is that a bargain to allow the project to proceed was not reached despite the offer of a reasonably large compensation package. Even when the utility changed the site to a location where there was quite strong local support, compensation that it offered at the new site did not result in an agreement. Bargaining is not just a matter of offering reasonably large compensation—it also involves taking into account competing industrial and political interests.

MOBILIZING SUPPORT

Chubu Electric publicly announced in early December 1963 that it planned to commence supplying power to the grid from the Ashihama plant in 1969. It set a seven-year developmental deadline allowing approximately two years to get community consent.[2] At that time, the company experienced a relatively low power reserve margin of 4 percent. Forecast demand for 1968 increased thirteen percent from the previous year. Supply capacity increased 19 percent. There was 2,720 megawatts (mw) of capacity in the licensing and construction stages, resulting in an expected supply surplus of approximately 552 mw.[3] The utility had sufficient capacity to meet expected demand increases, and there was no strong market justification for rapidly developing the project.

There was, however, a more pressing incentive in developing the project. In the early 1960s, Japan entered a nuclear energy boom, and there was intense competition among power companies for large government subsidies.[4] Chubu Electric's top management sought to secure a larger part of those subsidies, which traditionally went to the Tokyo and Kansai Electric Power Companies. This meant the Ministry of International Trade and Industry (MITI) had to be shown serious nuclear siting plans.[5] Getting national government subsidies for nuclear technology was more important than market justification.

Before its public announcement, the utility approached the Mie prefectural government to discuss a siting proposal. The governor, Tanaka Satoshi, previously a Ministry of Finance (MOF) official, had a strong development ethos. He believed a nuclear power plant would benefit Mie's economic development.[6] During the postwar period, Mie had been left behind neighboring areas, such as Kanagawa and Tokyo, but in the late 1950s entered a period of relatively high growth, which continued into the early 1960s. Tanaka also saw energy plants as a particularly attractive proposition because of Mie's proximity to Tokyo and Kanagawa and the prospects for exporting electricity to them. Within the community there was a strong desire for continued growth and little concern for issues of environmental quality.[7]

Tanaka ordered officials to investigate the Kumano region (Nagashima city and Kaiyama, Nanto, and Kisei towns) in southern Mie for potential nuclear power plant sites. This region was economically underdeveloped compared with the rest of the prefecture. Historically, it was prone to devastating typhoons. Continued reconstruction efforts were a drain on the prefecture's financial resources even during periods of solid growth.[8] Tanaka believed that the development of a nuclear plant in this region would free up resources to redistribute to more productive areas.[9]

In late November 1963, Tanaka held discussions with mayors on the feasibility of the proposal. They had little knowledge about nuclear technology, but all were concerned about the safety of the technology. On Chubu Electric's advice, Tanaka organized visits to the Tokai nuclear plant, the only other nuclear plant operating in Japan. The Tokai plant had a perfect safety record, and the four mayors were suitably convinced that the plant posed no risk to their local communities. They were also impressed with the huge economic benefits associated with the Tokai nuclear project. Tanaka moved quickly and persuaded them to respond by mid-December.

Tanaka's strategy was to induce competition among the four mayors to host the project.[10] There was considerable resistance from fishermen in Kaiyama town. It also became clear that Nagashima city was not suitable because of relatively high population density. In contrast, the conservative mayors of Kisei town, Yoshida Tamenari, and Nanto town, Nomura Junnosuke, eagerly supported the project.[11] These two mayors believed the project would benefit both towns. The site was located in a desolate area overlapping the boundaries of both towns, one of the few areas not suitable for agricultural production. The Liberal Democratic Party (LDP) had been the dominant party in Nanto and Kisei towns during the postwar period. Both towns relied heavily on external finances for the development of economic infrastructure. Yoshida and Nomura agreed to issue an investigation permit.

The utility viewed these developments positively. As one company official stated, "The site was ideal as far as we were concerned. Land was easily available. Elite leaders at both the prefectural and local levels were actively promoting the project behind-the-scenes to bring badly needed social and economic benefits to both towns. Both assemblies were dominated by conservative politicians. This meant that we did not foresee any ideological opposition emerging."[12]

Chubu Electric selected a site through an auctioning strategy conducted by the prefecture. It sought to gauge local reactions to the facility and judge which available sites were likely to yield the least political resistance or the most support. Those who clearly did not want a facility quickly indicated their opposition. Those who wanted to host a project actively encouraged it. This strategy aimed to minimize bargaining and compensation costs to the utility.

ECONOMIC RESISTANCE AND ACCESS TO POWER

With the agreement of both local governments, Chubu Electric commenced drilling investigations at Ashihama on 6 January 1964. At this stage there was an unexpected response from the local fishing industry in Nanto. In early March, the Kowaura fishing cooperative, the motivating force behind the emergence of opposition, established the Fishermen's Struggle Committee for Opposing Nuclear Energy. As shown in map 2, outlet pipes would discharge waste water into the area covered by their property rights.[13] There was a great deal of uncertainty about the effects of waste water on the marine environment, as one fishermen clearly highlighted: "While some fishermen were surprisingly not overly worried, the majority of us had images of marine life being cooked alive in a boiling ocean. We knew that the temperatures generated by anything nuclear were much higher than would normally be required to cook fish. Even though local officials told us otherwise, we did not believe them."[14]

The committee, realizing that the utility was determined to push ahead

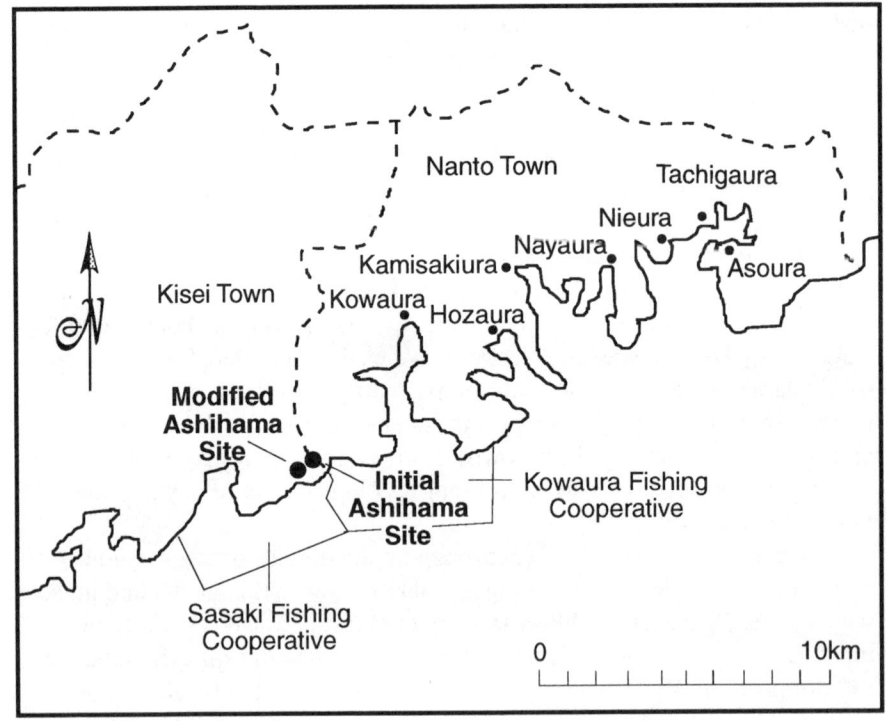

Map 2   Ashihama nuclear plant: site location and property rights ownership

with its plan, knew it had to mobilize stronger support for its position. It appealed to other fishing cooperatives in the Nanto region to oppose jointly. On 16 March, the Kowaura and six other cooperatives in Nanto (Asoura, Tachigaura, Nieura, Kamisaki, Naya, and Hozu cooperatives) successfully formed the Central Fishing Struggle Committee for Opposing Nuclear Energy (Central Committee).[15] This alliance formed the basis of resistance to the Ashihama nuclear plant.

The formation of the alliance resulted from fears and uncertainty about the effects of the project on the fishing industry. Yellowtail fishing had been the mainstay of the industry since before the Pacific War. By the late 1950s, a high incidence of typhoons reduced substantially the catch of yellowtail. The fishing industry shifted its efforts to pearl cultivation. The nearby coastline had several calm and shallow water inlets that provided an ideal environment for pearl cultivation.[16] In the early 1960s, there was a pearl boom in Nanto, and the value of pearl production increased to about 3,000 million yen per year.[17] The majority of the households in the region were employed in pearl production.[18] There was substantial concern about the impact of waste water. The fishing industry argued that temperature changes would affect stationary mollusc, such as pearls, more than fish, which were capable of migrating to other areas to escape temperature changes. The industry was also worried about possible employment consequences.[19]

The structure of the fishing industry facilitated alliance formation. Pearl cultivation was based on specialization in the production process. There were three major steps in the process, which took approximately three years: cultivating seedlings, which were attached to stationary nets; growing oyster shells; and inserting the seedlings into the oysters. The Naya fishing cooperative from Nanto produced the seed pearl, the five cooperatives from Kowaura to Tachigaura grew the oysters, and the Asoura fishing cooperative undertook the insertion process.[20] Given the integrated structure of the industry, fishing interests perceived that any adverse impact on the Kowaura fishing cooperative would affect negatively, and fairly evenly, the entire regional pearl industry. Furthermore, the uncertainties were spread relatively flatly across the whole industry. The gains (or losses foregone) from collective action on the part of the fishermen were very high.

Established historical relationships among the fishing cooperatives in the region also brought the alliance together. There was a long history of protecting their fishing grounds from incursions by outside fishermen. During the Tokugawa and Meiji periods, there were major conflicts between fishermen from Nanto and from Aichi prefecture over fishing rights in the waters off the Kumano coastline, which were richer in marine resources. Aichi fishermen periodically entered the Kumano area to fish. Nanto fishermen resisted these incursions, and several fierce sea battles occurred. There was a strong historical consciousness for collective defense against outsiders perceived to be damaging Nanto interests.[21]

The Central Committee's core strategy was to register its concerns through relevant political channels. At the prefectural level, it appealed to Satonaka Masahira, who was speaker of the prefectural assembly and who had fishing connections in Nanto. Satonaka was born in Nanto, his electoral support base. He consistently promoted the pearl industry in return for electoral support. As speaker of the Assembly, Satonaka shaped the nature of debate and exerted considerable influence on prefectural politics. The Central Committee used his influence to voice its concerns in the prefectural assembly.[22]

The Central Committee established its decision-making headquarters in the Mie prefectural fishing cooperative. The director of the cooperative was Ishikawa Maruyoshi, a powerful politician who had been a LDP prefectural assemblyman and a member of the House of Representatives. Ishikawa had close connections with prefectural politicians and exerted pressure on the prefectural government. He was also concerned that prefectural economic development policies were not adequately taking into account fishing interests. Backed by the Mie prefectural cooperative, he urged Governor Tanaka to withdraw his support for and to stop the project.[23]

Fishing interests had direct access to prefectural decision-making centers, which had an immediate impact. The proposed project caused a split within the LDP between rural interests wishing to protect the regional fishing industry and commercial interests that wanted to facilitate broader prefectural economic development. On 13 April 1964, Tanaka, addressing the nuclear issue in the prefectural assembly, declared the project would not be developed unless it could facilitate regional development and enhance the welfare of the regional community.[24] This guarantee meant enhancing the welfare of fishing interests. The politicization of the siting issue forced Tanaka to adopt a more cautious stance on the project while leaving open the option of developing or not developing the project.[25]

The Central committee also appealed to the Nanto town assembly. In particular, it sought to recruit to their cause assemblymen representing the farming sector. This strategy was critical in changing the assembly's neutral position on the project.[26] In Nanto, LDP assemblymen represented fishing, farming, and commercial interests. Historically, local politics was dominated by farmers. With the emergence of the pearl industry, the fishing sector became a new political force in the town. The agricultural sector, despite a relative economic decline, nonetheless still retained considerable influence in local politics. The small commercial sector, represented by Mayor Nomura, had only minor representation.

The changing local industrial structure was accompanied by conflict between agricultural and fishing interests over the spoils of government. Nomura came to power in the late 1950s, a time when this conflict was particularly intense, because of his ability to mediate between fishing and farming interests. During the initial stages of the siting conflict, Nanto farmers adopted a neutral position. The continued resistance by fishermen, however, persuaded them that the nature and extent of benefits from the project were unclear. They also became concerned about the risk of nuclear energy and the potentially harmful effects on their agricultural produce. Opposition by fishing interests increased uncertainty about the value of the project on the part of farming interests.

Fishing industry appeals were successful.[27] On 17 June 1964, the town assembly decided to oppose the project. Undeterred, Nomura continued to promote the nuclear plant. In response, the Central Committee together with the farming sector started a recall movement in late July, and Nomura was ousted as mayor of Nanto. On 30 August, after discussions between fishing and farming interests, Yamamoto Tenzo, a farmer from Naya hamlet, came to power in an uncontested election. Although Yamamoto was a farmer, all of the assemblymen in his assembly were fishermen. The Central Committee believed its ability to oppose would be weakened should be farmer–fisherman conflict reemerge in the context of selecting a suitable mayor. The division of power represented a compromise between fishing and farming politicians. It closed political channels via Nomura, which had been open to both the utility and the prefecture—both of whom relied too much on the local mayor to facilitate decision-making. The Central Committee, having gained complete political control over the town assembly, was in a powerful position to oppose the project.

Although auctioning strategies were useful in mobilizing elite leadership support for the project, they were not effective as elites did not take into account the project's effects on regional interest groups. Elite responses gave a incorrect impression of the bargaining and the compensation costs involved in the siting. The costs to regional interest groups were much higher than those judged by elites. They underestimated significantly the intensity of opposition and the bargaining and compensation costs involved in reaching an agreement. The auctioning process was a closed one and was not structured to generate pluralist responses.

Fishing interest groups opposed effectively and increased the bargaining costs to private and prefectural promoters. These groups formed an exclusive coalition and gained access to local political power. A fairly even spread of adverse impacts across the fishing industry, high uncertainty about the expected costs of the project, and historical experience in resisting unwanted outsiders facilitated coalition formation and collective action. The coalition succeeded in altering the distribution of power in the bargaining environment. It appealed to both prefectural and local assemblies. Most important, it increased uncertainty about the costs of the project to other local interests and skillfully managed intra-party conflict to gain access to local government. Opposing unwanted projects requires both collective action and using that action skillfully to alter the distribution of political power.

POLITICAL FACTIONALISM

Surprisingly, on 30 July 1964, soon after the Central Committee started the recall movement to oust Nomura, Chubu Electric consulted Governor Tanaka and formally decided on the Ashihama site. On the same day, the Kisei town assembly accepted the proposal.[28] The economic incentive to accept the proposal was much stronger in Kisei than it was in Nanto. Unlike Nanto, which had a very lucrative pearl industry, Kisei had no major industry to sustain local development. Local leaders viewed the project as a way of stimulating the local economy. Paradoxically, however, it was precisely because the project was so appealing that it was delayed, as competing factions in the local LDP squabbled over credit for the political gains associated with the development.[29]

Kisei had been formed in 1957 by an amalgamation of Sasaki town and Kashinozaki town. In contrast to Nanto, the Sasaki town economy was based on a small deep-sea fishing industry.[30] The development of the industry, with its need for larger boats, a port, and processing facilities, left the Sasaki fishing cooperative with large debts. Since Sasaki was no more than a base for fishing operations conducted at some distance, members believed the project would not adversely affect their industry. There was significantly less uncertainty than in Nanto. Indeed, deep-sea fishermen saw the sale of coastal property rights as a means of relieving their financial problems.

Kashinozaki town, located in a mountainous inland area, had even fewer reasons to object to the project. The major industry, forestry, had suffered a general decline during the postwar period and was not capable of sustaining local development. A severe lack of infrastructure, such as roads, posed a major problem for the mountain town. Local citizens saw the construction of the project as very positive, particularly given relatively high income growth in neighboring Nanto.

Yoshida Tamenari, the mayor of Kisei, used these economic incentives to promote the nuclear plant. He was, however, constrained in his support because of an opposition faction within the local LDP headed by Sakaguchi Yuzo, the speaker in the Kisei town assembly. The Sakaguchi faction argued that Yoshida would be able to strengthen his power base in the town should the project go ahead. As an astute observer remarked, "The conflict between Yoshida and Sakaguchi was not over the safety of the nuclear power plant even though, on the surface, several concerns about the impact of radiation were raised. Rather, it was over intra-party competition and the acquisition of political power within Kisei town."[31] To a certain extent, this conflict delayed assembly approval of the project.

The origins of the Yoshida and Sakaguchi factions lay in the pre-amalgamation days. In the early 1950s, the national government had encouraged local amalgamations to enhance the administrative efficiency of local governments. In 1953, Mie prefecture decided that Sasaki town would amalgamate with Nagashima town and that Kashinozaki town would amalgamate with Ouchiyama town. The rationale was that the former two towns were mainly fishing economies, whereas the latter two were forestry and farming ones.[32] The ruling Sakaguchi faction supported the amalgamation plan, but the Yoshida faction opposed it successfully.

Yoshida's father, a yellowtail fisherman, had considerable influence within Sasaki. Fearing that an amalgamation with Nagashima would weaken his father's influence, he opposed Sakaguchi. He argued that the Sasaki assembly would be absorbed into the Nagashima assembly, that many assemblymen would lose their positions of influence, and that the Sasaki fishing cooperative would lose its fishing rights to the bigger and more powerful Nagashima fishing cooperative. These arguments created opposition among assemblymen and fishermen to Sakaguchi's plan for amalgamating with Nagashima. At the same time, Ouchiyama rejected the prefecture's desire for amalgamation with Kashinozaki. It represented a financial burden that the Ouchiyama assembly rejected. Yoshida saw that Kashinozaki faced a problem in trying to find an amalgamation partner. He persuaded his father to propose amalgamating with Sasaki. Kashinozaki assemblymen overwhelmingly accepted this proposal as they judged their influence would not be jeopardized by uniting with the smaller Kashinozaki town.

After the amalgamation in 1957, conflict emerged within the Sakaguchi faction over leadership in the coming 1959 Kisei town election. There were two other influential people in the Sakaguchi faction: Taniguchi Tomomi, deputy speaker of the Sasaki town assembly and a powerful figure in the fishing cooperative, and Nakaseko Bunji, head of the cooperative. Sakaguchi, Taniguchi, and Nakaseko had been primary school classmates. Both Taniguchi and Sakaguchi had leadership ambitions. Taniguchi was in conflict with Nakaseko over the leadership of the Sasaki fishing cooperative. Taniguchi, a large boat owner who often laid off fishermen during recessed periods, was not popular among fishermen and joined the Yoshida faction. Yoshida accepted the support of Taniguchi. Nakaseko, fearing that

Taniguchi's position in the Sasaki fishing cooperative would be enhanced, decided to back Sakaguchi. In the 1959 election, the Yoshida-Taniguchi faction came to power.

In this way, the ruling Yoshida-Taniguchi and opposing Sakaguchi-Nakaseko factions were formed in Kisei. The Sasaki fishing cooperative was a major source of political support to both Yoshida and Sakaguchi. Neither could formulate or implement town policy without the consent of either Taniguchi or Nakaseko, whose core electoral support came from the Sasaki fishing cooperative.[33]

The Sasaki fishing cooperative, under the leadership of Taniguchi, announced a policy of opposing the project on 17 June 1964, approximately a month before the Kisei town assembly accepted it.[34] Sakaguchi and Nakaseko were not capable of preventing the assembly decision, as Yoshida obtained support from Kashinozaki assemblymen, giving him the numbers to obtain assembly approval. The balance of power in Kisei lay with assemblymen from Kashinozaki, who had eight members in the town assembly. Yoshida was able to persuade them to accept the project. Despite political resistance by the Sakaguchi-Nakaseko faction, he argued that its development would bring social and economic benefits to the Kashinozaki area.

Intra-party conflict between ruling and opposing conservative factions over the project siting emerged. The project had significant political and economic impacts on the small local community. It became a lightning rod for factional conflict as there was uncertainty about who would gain and who would lose politically and economically. Although this intra-party conflict caused some political problems, it did not significantly delay key assembly decisions. The leadership of the ruling faction struck an implicit deal with the opposition that gains from the project would be shared relatively equally among competing constituency groups.

REVISED EXPECTATIONS AND PROMOTER RESPONSES

Meanwhile, changes in the electricity supply strengthened the market need for the project. Between 1965 and 1967, Chubu Electric's expected five-year shortfall increased dramatically from 163 mw to 885 mw.[35] Expectations of expanded economic activity, particularly in Mie, Shizuoka, and Aichi prefectures, led to a revision of projected electricity demand trends. The project's original objective was to compete for subsidies from the national government. This goal then became overshadowed by the need to avert electricity shortages. The power company revised its completion schedule. In July 1965, despite continuing resistance by the Nanto assembly, it announced that it wanted to reach a settlement within a year so that construction could commence by early 1967.[36]

The utility was hesitant to bargain directly with the Nanto cooperative over the transfer of fishing rights. Based on intelligence it had received, it judged the cooperative would demand at least 3,000 million yen in compensation for the exchange of property rights. The cooperative was in a powerful bargaining position. It not only had alternatives to negotiating a siting deal but also had the political power to pursue those alternatives. The utility was only willing to pay approximately 900 million yen.[37] It judged that 3,000 million yen would increase construction costs by approximately ten percent and reduce the economic viability of the project, even with government subsidies.

Interestingly, other power companies also contributed to Chubu Electric's hesitancy to offer large amounts of compensation. Because the Ashihama project was one of the first nuclear plants in Japan, other utilities were worried that the payment of exorbitant compensation might become a precedent for other siting disputes, thereby creating a trend detrimental to the development of the nuclear industry.[38] An official from Chubu Electric summed up this concern as follows:

> While we had a tough time with the local fishing industry, it seemed as if we had a tougher time with other power companies, which were all eagerly promoting nuclear power. They were telling us that we should not offer large compensation packages to fishing cooperatives because it might affect planning for their current and future projects. While we did not like other companies telling us how to manage disputes in our electricity sphere, we appreciated their concerns given the "similar situation" clause in the Compensation Standards.[39]

Pressure from other utilities because of possible negative spillover effects on concurrent and future siting negotiations at their sites constrained Chubu Electric's options in negotiating with fishing interests.

Chubu Electric could not rely on substantial support from MITI in reaching a settlement. The market situation in the central electricity sphere did not justify speedy development of the project. From 1963 to 1967, the expected electricity shortage declined from 6,833 mw to 1,896 mw.[40] MITI took the view that the utility could purchase electricity from Kansai Electric if the situation warranted it.[41] It also believed that intervention might intensify local resistance.[42]

Constrained by other utilities and unable to get support from MITI, Chubu Electric relied more heavily on the prefectural government to broker a settlement. The prefecture had played an important role in selecting the Ashihama site. The company persuaded Tanaka to adopt a more positive attitude toward the project. Although he had lost some political support at the prefectural level, he thought he could regain that support if he could mediate an acceptable agreement. Tanaka also knew that failure would jeopardize his chances in the 1967 election, which was only two years away. He felt that if he were going to take a risk, then he would have to maneuver quickly to break the impasse.[43]

The prefecture, in cooperation with Chubu Electric, devised a two-tier strategy for changing the attitude of the Central Committee. The first step involved the prefectural government commissioning an investigation into the impact of waste water on the marine environment. The aim was to reduce uncertainty about the expected costs to fishing interests. Marine experts commenced the investigation in October 1965 and in December that year published an interim report. The report was favorable to the utility's position, stating that waste water discharges would increase the ocean temperature only in the vicinity of the outlet pipes by a small amount and that this would not adversely affect the fishing industry. It concluded that the prefecture should discuss the results with the Central Committee to allow further investigations.

The Central Committee flatly rejected any meetings with prefectural officials. It issued a declaration that criticized the report. First, the researchers had been selected personally by Tanaka and had conducted the investigation on the presumption that the project would proceed. Second, it was impossible to obtain even intermediate results in a two-month period. Any investigation should take at least a year, on the grounds that the impact of waste water might differ seasonally. Third, the report did not consider specifically the effects of waste water on pearl cultivation, which was critical to the Nanto economy. Any investigation ignoring pearl cultivation would not be accepted by the Central Committee.[44]

The second component aimed to provide a prefectural subsidy for regional development. On 15 November 1965, the prefecture published a plan, which incorporated a direct subsidy amounting to 6,700 million yen, for the provision of infrastructure and for the development of the fishing industries in Nanto, Kisei, and Nagashima.[45] The prefecture hoped that such benefits, which were actually similar to those provided for under the Three Laws established later in 1974, would intensify local and regional support for the project and weaken the Central Committee.[46] Kisei and Nagashima were willing to accept the subsidy, but the Nanto assembly rejected it. It claimed the subsidy, though providing broader regional benefits, did not provide any guarantee of direct benefits to fishing interests.

The Nanto assembly responded decisively to this economic pressure by strengthening the organizational and financial structure of the opposition movement. On 20 November 1965, it created the Nanto Strategy Committee for Opposing the Nuclear Power Plant, which comprised all local assemblymen. It decided on a budget, drawn solely from town funds, to finance a petition. By the middle of December, it had compiled a petition with 8,023 signatures, which it sent to the prefecture stating Nanto opposition to the Ashihama nuclear project.[47] The attempt by the prefecture to mediate in the conflict, by reducing uncertainty about environmental costs and increasing economic benefits through direct subsidies, was highly innovative but only intensified opposition.

The continued resistance in Nanto forced Tanaka again to adopt a very cautious posture. The political cost of further pressure from the prefecture had become high. The assembly was split over the issue. The 1967 prefectural election was now only a year away, and Tanaka believed that further conflict between the prefecture and Nanto town would have significant electoral implications. Despite its enthusiasm in principle for the project, the prefecture now sought to end it. On 22 May 1966, Chubu Electric announced its intention to commence investigations necessary for Electric Power Development Coordination Committee (EPDCC) approval and to complete those investigations by June. The prefecture, knowing that any action by the utility was politically risky, tried to persuade it to delay investigations.

Changed economic circumstances led to revised expectations about the expected benefits of the project, which encouraged the utility to try to accelerate bargaining processes. Although these changed expectations increased the compensation pool, they did not enable promoters to break the gridlock. The pool was not large enough to allow promoters to gain consent from interest groups without significantly reducing the project's economic viability. Other developers were worried about the negative implications of large compensation outlays on their planning processes. Revised expectations did not alter fundamentally the bargaining environment.

The utility brought in the prefectural government as a third party to assist in mediating the dispute. The aim was to increase the compensation pool available to adversely affected interests. This change in the pool was to be accomplished by providing information that would reduce uncertainty about perceived risks and increase the benefits by offering subsidies. Third-party intervention did not work. The prefecture had lost any legitimacy it had by its approach in conducting the environmental investigation. Furthermore, the subsidy plan only provided indirect benefits to key political players with veto power, even though it provided direct benefits to the regional community as a whole. The prefecture's approach intensified opposition. It could not develop a workable conduit into local political processes and get support from local politicians to play a mediatory role.

RISK MITIGATION AND INTRA-PARTY COMPETITION

Thwarted even at the prefectural level where political support was the strongest, Chubu Electric turned to the local level in an attempt to weaken the Central Committee politically. Undaunted by the intensifying opposition, it commenced construction preparations in Kisei. It also decided to seek support from Yoshida to change the design of the project and to persuade national politicians to enter the conflict and exert their political muscle. The utility hoped that this strategy, which bypassed the prefecture, would weaken the Central Committee's bargaining strength and create the necessary political conditions to strike an agreement.

In September 1966, the utility decided to change the proposed placement of the containment vessel from the boundary of Nanto and Kisei to solely within Kisei and to relocate the waste water outlet pipes to a position in Kisei. The aim of these changes was to reduce the bargaining power of those who opposed the project, and to increase the negotiating power of those who supported it. Chubu Electric believed that these design changes would nullify the need to relinquish property rights possessed by Nanto fishing cooperatives. They would only have to secure the property rights from the Sasaki fishing cooperative, which supported the project given its deep-sea fishing operations and its poor financial situation. The power company hoped it would only be expected to pay "cooperation money" to Nanto fishing cooperatives. This strategy also made sense to the utility as it had developed good relations with the Yoshida-Taniguchi faction in Kisei, and it was through this faction that Chubu Electric attempted to execute the strategy.

Yoshida was delighted with the design change, as he knew it would bring significantly more benefits to Kisei. With the containment vessel solely in Kisei, it would become the exclusive recipient of fixed-assets taxes, instead of having to share them with Nanto. He thought these increased financial benefits would enhance his political position vis-à-vis Sakaguchi and persuaded Taniguchi to allow full-scale investigations to start. In late August 1966, Yoshida agreed to issue the utility an investigation permit.

Chubu Electric also enlisted the support of influential national politicians to persuade the Central Committee that the project was in the national interest in terms of catching up with the West technologically. In September, the House of Representatives Science and Technology Promotion Policy Committee, chaired by Nakasone Yasuhiro of the LDP, decided to visit the Nanto area. Whereas MITI was not willing to enter into the regional conflict, Nakasone had a particular interest in the development of the Ashihama plant. He was the architect of Japan's decision to introduce nuclear power and wanted to see nuclear development proceed as quickly as possible.

The Central Committee was concerned about the company entering the local area and conducting investigations. Members developed contingency plans to stop any incursions. They conducted training exercises to stop vessels entering the area and undertook daily surveillance operations off the Kumano coastline. The committee established a policy of dealing with the entry of foreign vessels into the area. As one leader of the Central Committee explained vividly,

> They [Central Committee] created seven mobilization groups and called them the Seven Samurai.[48] Each group had a specific function. For example, one group monitored foreign vessels in particular areas, another relayed information about movements, and another intercepted those vessels. Given the intensity of feelings, leaders were very concerned that any action might get out of hand and that some members may resort to violence. To minimize this possibility, they established rules that stipulated that a decision to intercept a vessel had to be made by the chairman of the Central Committee and that violence could not be used under any circumstances.[49]

The committee mobilized the force on 18 September 1966 in response to a planned visit by Nakasone and other politicians. The visit resulted in the Nagashima Incident, the first and largest demonstration conducted by fishermen in the postwar period. Hundreds of fishing boats from Nanto surrounded and stopped a Maritime Defense Agency patrol boat off Nagashima city. Members of the Central Committee boarded the patrol boat to reaffirm their opposition.[50] Although several fishermen were arrested for unlawfully boarding the vessel, national politicians failed in their attempt to exert political pressure on the Nanto fishing interests.

Despite the incident, the power company continued its attempts to develop the project. On 15 November, Kisei town signed an agreement with Chubu Electric to allow it to commence full-scale investigations. Soon after, rumors emerged that the power company was giving financial contributions to Kisei town. These donations were apparently channeled to Taniguchi so that he could persuade fishermen in the Nakaseko faction to accept the project. During the latter part of December and the beginning of January 1967, these rumors became more widespread. Residents, particularly in Kashinozaki, were upset that Yoshida, by transferring contributions to the Sasaki fishing cooperative through Taniguchi, was promoting the project for the good of Sasaki at their expense.

Earlier expectations that the project would benefit the whole local community were abruptly shattered. With the project now located closer to them, Kashinozaki residents became uncertain about benefits from the project. They started a movement to receive compensation from the town and Chubu Electric. The utility was not prepared to entertain their requests. On 24 January, eight Kashinozaki assemblymen resigned from the Kisei assembly. The Sakaguchi–Nakaseko faction, seeing this as an opportunity to seize power, joined forces with the Kashinozaki assemblymen, and began a recall movement. Yoshida lost the vote decisively by 1,344 to 2,242 votes, and in April Sakaguchi was elected mayor. Reflecting the changed allocation of power, Sakaguchi immediately announced a policy of opposing the project.[51]

Chubu Electric was then confronted with opposition from both the Nanto and Kisei town assemblies. The prefecture was no longer willing to support the project. The national government was hamstrung. The utility lost any direct link to key local power brokers and could not rely on third-party mediation at any jurisdictional level to resurrect bargaining processes. Having already changed the location of the plant to Kisei, it was locked into that site and could not seek politically to relocate it back at Nanto. It decided to abandon the project and consider an alternative location outside the Kumano area on which to site its first nuclear power plant.

In seeking to break the impasse, the utility changed the location of the site (the ultimate form of design modification) to facilitate local bargaining processes. This strategy sought to alter the structure of the bargaining environment by weakening the power of Nanto fishing interests and enhancing the bargaining positions of supporters in Kisei. It judged that Kisei's bargaining power was weaker than Nanto's, as the former had relatively few alternatives to striking an agreement. With the site change, the company certainly had adequate compensation to strike a deal with the poorer Kisei community. As in Nanto, however, it relied too much on the ruling faction in its approach to siting negotiations. It did not structure its compensation policy to take into account intra-party competition. When rumors of bribery emerged, locals supportive of the nonruling faction changed their expectations about the worth of the project. This led to a political backlash from which the company could not recover.

The use of political pressure from the top was counterproductive in breaking the gridlock between the utility and fishing interests. These interests not only had adequate alternatives to striking a bargaining but also had the political mandate to back that up. In this case, not even national political pressure could weaken the resolve of local interests. Indeed, it actually intensified opposition and created a situation in which any sort of political compromise became impossible. The case shows how local communities that do not want to accept unwanted projects can refuse them, even in the face of intense political and economic pressure from extra-local interests.

CONCLUSIONS

Ashihama highlights a classic NIMBY case, a developer's worst nightmare, where the use of compensation was ineffective in negotiating an agreement over an unwanted facility. Promoters offered compensation for property rights transfer (approximately 900 million yen), significant prefectural subsidies for regional development (6,700 million yen), considerable risk mitigation through the use of the demonstration effect, expert investigations, a major design change, and even alleged under-the-table money. Although similar compensation packages resulted in agreements in other cases, they did not work in the Ashihama dispute.

The structure of the bargaining environment was shaped in a way that, according to the statistical analysis, would have resulted in a slightly longer settlement time (eighty-seven months) than the average of initial nuclear plants (eighty-two months). Chubu Electric faced moderately large electricity shortages during the dispute, and it wished to develop a plant quickly. Electricity shortages were declining and the national government placed less emphasis on rapid implementation. Prefectural incomes were growing quickly and the community wanted that growth to continue. Economic opportunities in the rural sector were very high, and there were grave concerns about environmental harm. The financial ability of local government was limited, and there was a strong interest in the project's economic benefits. There was little Leftist party influence at the local level, and community attitudes placed little emphasis on environmental quality.

The polimetric models provide an important perspective for understanding the structure of the negotiating environment in which this dispute was played out. The utility clearly sought to strike a quick bargain. Local and prefectural governments provided strong support, particularly in the early stages. Regional fishing interests saw the costs as being quite high and represented the major opposition force. Taken together, these results suggest a bargaining environment where an economic surplus existed and where promoters should have been able to redistribute some of their expected gains to rural interests to resolve the dispute. Collectively promoters offered a reasonably large compensation package.

Although these insights are important, the models were much more inconsistent with actual outcomes than the other case histories examined in this book. Furthermore, they did not reveal that the project would be ultimately abandoned. This divergence can be explained by the ability of opposing interests to gain electoral access to political power. Although opponents in siting conflicts influence local decision making in different ways, they were able to dominate local assemblies in the Ashihama case. Promoters employed a range of bargaining and compensation strategies but could not break the impasse.

The major costs of the project fell heavily on fishing interests. Given the pearl boom in Nanto, the environmental risks of the project were perceived to be large. The structure of the industry was characterized by close historical relationships and production specialization. As the costs of the project were spread evenly across the cooperatives in the industry, they were able to engage in effective collective action. The per-capita costs of not engaging (or the benefits from engaging) in collective resistance to each of the interests were very high, which mitigated the free-rider problem commonly associated with collective action. The opposition was able to intensify resistance at different stages of the dispute in response to the actions of promoters.

The structure of bargaining power ultimately favored local interests. Fishing cooperatives in Nanto clearly had viable alternatives to negotiating a settlement. Even with compensation offers, a course of action for the cooperatives, which maintained their existing emphasis on pearl cultivation, was more attractive. In contrast, local interests in Kisei were initially in weaker bargaining positions. They judged that their existing developmental options were less favorable than the alternative option, which included the project. This judgment explains the relative lack of opposition by both community and fishing interests during most of the dispute. When bribery allegations arose, some local political groups that felt they were going to be disadvantaged decided that the alternatives to having the project were much better. Communities in both towns, given their preferred alternatives and their access to local assemblies, were able to increase the negotiating costs to the point where the utility decided that an alternative site was less costly.

Opposing interests were able to counter promoter bargaining strategies during the course of the dispute. Proponents used a range of strategies, including intermediation by third parties, offering subsidies, and risk mitigation to get fishing interests to the negotiating table. When these did not work, they employed overt political pressure and bribery tactics. Opponents countered these tactics effectively, appealing to relevant political players in both local and prefectural arenas, stimulating local intra-party competition, and skillfully gaining access to local political power. They prevented attempts at creating negotiating tables (Nanto) as well as dismantling those tables (Kisei). Local interests handled a barrage of coordinated strategies from powerful extra-local economic and political interests.

Expectations about the value of the project changed during the bargaining process and ultimately pushed the conflicting parties further away from any willingness to compromise. Revised expectations arose mainly from endogenously induced changes. Continued successful resistance by Nanto fishing interests changed the utilitys' expectations and led to the site change. Promoters' strategies, particularly those relating to the distribution of compensation, changed local expectations in Kisei about the value of the project. Like their Nanto counterparts, they gained access to local power. Expectations changed in a way that increased negotiating costs for the conflicting parties to the point where abandoning negotiations became preferable to both sides.

High levels of uncertainty characterized the dispute. Proponents sought to manage that by reducing the costs and increasing the benefits of the project. But the way they did this, particularly through ill-perceived, expert investigations, inadequate compensation offers, and the use of pressure and force, enhanced local uncertainty not only about the value of the project but also about promoter motivations and interests. The lack of perceived legitimacy of those strategies exacerbated already existing local uncertainties. Opponents gained access to local political power to manage their uncertainties, which further increased promoter hesitancy. The power company judged that the likely time horizon for resurrecting bargaining made it preferable to explore alternative sites.

Bargaining can end up in stalemate in Japan as the Ashihama and other failed attempts to site noxious facilities illustrate, particularly where local interests alter fundamentally the structure of the bargaining environment by gaining access to local political power. Yet there are many more cases in which local interests are less powerful and where conflicting parties, while starting from widely diverging positions, can negotiate compromises even if it takes long periods of time to do so. Ashihama provides a useful benchmark against which to consider the next three disputes, in which agreements to construct energy facilities were reached.

CHAPTER FIVE

Carving Up Opposition Alliances

Negotiating siting outcomes requires skillful execution of myriad time-consuming bargaining strategies. It is not just a matter of deciding who needs be compensated and by how much and in what forms. A critical objective of any bargaining process is to create a suitable bargaining table on which compromises can be agreed. In bargaining, it is important to keep open the doors to negotiation or to open them where they are closed. Bargaining processes and outcomes will be influenced by the extent to which promoters manage effectively distributional issues in ways that minimize unwanted interference.

Hamaoka is important in that it contradicts the notion that NIMBY disputes take a long time, if ever, to resolve. It took just under two years to reach an agreement and was one of the fastest settlements in the history of nuclear power sitings in Japan.[1] Although we need to know why some projects are either abandoned (Ashihama) or delayed for longer periods of time (Matsushima and Tomari), we also need to know why others are sited much more easily. By comparing Hamaoka with Ashihama, we can start to explore the conditions under which compensation is likely to be an effective tool in resolving siting conflicts.

The Hamaoka case is fascinating in contrast with other cases because only very small amounts of compensation were required to reach a settlement with community interests. Fishing cooperatives opposed the development and formed a coalition with Leftist political interests. Regional power brokers split the alliance, isolated this ideological interference, and negotiated a settlement with little compensation. Even though resistance was weaker than at Ashihama, there were conditions that could have resulted in another abandonment had the utility not considered the lessons it learned from its experience at Ashihama.

ECONOMIC AND POLITICAL SUPPORT

In July 1967, Chubu Electric made a decision to locate a nuclear plant in Hamaoka. It anticipated that it would receive Electric Power Development Coordination Council (EPDCC) approval by December of that year.[2] After the Ashihama project was abandoned, Chubu Electric's siting options for large projects appeared to be restricted to the Shizuoka prefecture. The company supplied electricity to the Nagano, Gifu, Aichi, and Mie prefectures in addition to Shizuoka. Siting nuclear and conventional plants was not feasible in Nagano and Gifu because of mountainous terrain and a lack of adequate cooling water. Few sites remained for large hydroelectric capacity. Pollution problems in Aichi were hampering the siting of a host of energy and industrial projects in that prefecture. Attempting to site any plants in Mie was politically risky, if not impossible, given continued sensitivity over the Ashihama dispute.[3]

The speed at which the company wanted to site the plant was directly attributable to an expected electricity supply shortage. Between 1966 and 1967, expected electricity demand calculated in the utility's five-year forecast increased steadily from 4,950 megawatts (mw) to 7,840 mw. Actual supply increased from 4,950 mw to 5,345 mw. The utility was relying on Ashihama and the Nishi-Nagoya and Atsumi fossil-fueled plants, which were under construction, to fill about half of the expected shortage. The abandonment of the Ashihama plant placed the company in a position of expecting a shortfall of approximately 1,900 mw in supply capacity by 1972. The cost of not successfully siting the Hamaoka project was very high.

An additional worry was the expected supply situation in the central electricity sphere comprising the Chubu, Kansai, and Hokuriku power companies. Between 1967 and 1969, expected shortages increased at 26 percent per annum. The Ministry of International Trade and Industry (MITI) monitored the situation with concern and valued the Hamaoka project highly in terms of balancing the electricity market in the central sphere. Chubu Electric was also concerned that it might not be able to purchase electricity from other utilities if the need arose. The priority it placed on the project influenced the speed at which it wanted to obtain community agreement and the willingness of MITI to lend its support.

In early 1967, Kato Osaburo, president of the company, discussed the proposal with Mizuno Shigeru, president of the *Sankei shinbun,* and an influential businessman in Shizuoka.[4] Mizuno was very enthusiastic and discussed the matter with Maruo Kenji, a prefectural Liberal Democratic Party (LDP) assemblyman. Mizuno and Maruo both had been born in Hamaoka in the Sakura hamlet and were close friends. They believed that Sakura would be an ideal site.[5] Both cultivated an interest in the project at the prefectural level.

The prefectural government published the seventh economic development plan in 1966. The nuclear project was consistent with achieving policy priorities embodied in the plan.[6] The first priority was to sustain high levels of economic growth. During the immediate postwar period, Shizuoka lagged behind the rest of Japan. In the late 1960s, it entered a high growth period. Kato found it easy to persuade the prefecture to accept the nuclear plant to further stimulate economic development. The prefectural government was not greatly concerned about the environmental implications of the project.

The second priority was to increase electricity self-sufficiency. Prefectural electricity demand was increasing at a relatively fast rate. Hydroelectric development, the major source of electricity supply, was constrained by a lack of suitable sites. During the early 1960s, Shizuoka turned from being a net exporter to a net importer of electricity and became more reliant on electricity generated from energy plants in Tokyo and Kanagawa.[7] As electricity demand in those areas was increasing rapidly, the prefectural government became increasingly nervous about relying on outside electricity sources.[8]

The third priority was to promote balanced regional economic development. Shizuoka was divided economically into the eastern, western, and southern regions. The eastern and western regions were experiencing rapid growth. They were geographically close to the Keihin and Kansai districts, which were registering strong growth in demand for goods and services. The more isolated southern region lagged behind the rest of the prefecture. MITI designated the region comprising the Hamaoka, Omaezaki, and Sagara towns and Yoshida city as underdeveloped in 1966.[9] The prefecture was particularly interested in stimulating growth and development in this region.

The power company, through Maruo, established connections with Kamogawa Tadaichi, an important figure with substantial behind-the-scenes clout in Hamaoka. Kamogawa, like Maruo and Mizuno, was born in Sakura hamlet and was an influential landowner before the Pacific War. With the backing of the agricultural sector, he became the inaugural mayor of Hamaoka in 1955 and was head of the Sakura agricultural cooperative. During the 1960s, the cooperative had experienced declining revenues. Kamogawa had a strong interest in selling land for the project development and saw the siting as a way of managing the cooperative's financial problems.[10]

Kamogawa also thought the project would stimulate the town economy by providing a stronger tax base and an expansion of employment opportunities. Hamaoka, like other areas in the southern part of Shizuoka, was economically underdeveloped. Approximately 60 percent of the workforce was employed in agriculture, with a large proportion being part-time farmers. The town had no industrial base. Many young people were leaving in search of better employment opportunities in urban areas. The town relied heavily on prefectural and national government finance for infrastructure.[11] In contrast to Nanto, Hamaoka had fewer developmental alternatives, which shaped its bargaining position.

In discussions with utility officials, Kamogawa suggested keeping a tight lid on the proposal until after local elections scheduled for March 1967. This secretiveness was not because of any perceived Leftist strength in the local assembly. It was because of deep dissatisfaction within the LDP with Mayor Shinozaki Tadaichi's economic policies. He anticipated Shinozaki would use the siting issue to regain support from other factions. Kamogawa wanted to get rid of Shinozaki and to have a mayor elected from his own faction. He assessed that raising the nuclear issue before the election would lead to party infighting, which might jeopardize his election strategy and the future of the project.[12]

Chubu Electric accepted this advice without hesitation. Even though the market justification for the Hamaoka project was strong, it delayed any further negotiations to avoid the project getting entangled in intra-party politics. The bitter memory of intra-party conflict at Ashihama was still fresh in its mind. Ashihama taught the company that pushing too quickly for a settlement where factional conflict existed could generate substantial opposition, even where economic imperatives were conducive to community acceptance of projects.

Kamogawa successfully supported Kawarazaki Mitsugi in the election. Kawarazaki was also born in Sakura and was a school colleague of Maruo. Kamogawa believed that Kawarazaki would support the proposal:

> Kamogawa called to congratulate me on my election win. He told me of his and Maruo's plan to develop a nuclear project in Hamaoka. I was stunned. They really kept the plan under wraps. I immediately spoke to Maruo, who convinced me that the project would be in the best interests of the town and prefectural economies. Maruo pledged his total support and backing. I was very nervous about raising the issue with the electorate at that stage and knew very little about nuclear power technology. But I had to proceed given the backing he had given me in the election.[13]

Kawarazaki immediately established the Hamaoka Development Investigation Committee (Investigation Committee), which comprised town politicians and officials, and provided political and administrative support for the project.[14]

The skill with which developers approach local communities has important implications for bargaining. In contrast to Ashihama, where the utility approached elected officials, in Hamaoka it approached behind-the-scenes power brokers in gaining support. This approach allowed brokers to manage intra-party conflict, which could have increased political uncertainty. The utility learned at Ashihama that siting did not involve simply deciding on a site, getting elected officials to announce the decision, and then defending it. Chubu Electric was more politically astute this time. It realized that real political power often rested with behind-the-scenes political operators. It identified these power brokers and encouraged them to provide the overall strategic and tactical input in the siting process.

EMERGENCE OF REGIONAL OPPOSITION

Chubu Electric's siting plan was leaked to the *Sankei shinbun* on 5 July 1967.[15] Opposition by Leftist political parties and their affiliates, local residents in Hamaoka, and regional fishing cooperatives rapidly emerged. By October, they had formed an alliance that, although financially weak, geographically dispersed, and ideologically diverse, was in a position to jeopardize the project. Promoters did not expect the scope of the opposition that emerged or the speed at which it developed an organizational base. Their lack of preparedness led the utility to underestimate the time required to win community approval for the project.

The Japan Socialist Party (JSP) and the Japan Communist Party (JCP) quickly became vocal on the issue not only in Hamaoka but also in Ogase city. They were concerned about two safety issues. The first was that Hamaoka was relatively more densely populated than other areas, such as Fukushima and Mihama towns, where nuclear power plants were also being proposed. The second, related issue was that a fault line ran to the north of Hamaoka, and there were doubts about the ability of the containment vessel (which houses the nuclear reactor) to withstand a large earthquake.[16] One member of the anti-nuclear movement highlighted this concern:

> We were concerned that nuclear technology was a relatively new technology and that it should not be introduced commercially until its safety was verified. But what we really were concerned with was that the Hamaoka plant was risky. It was the closest nuclear plant to Tokyo. If there were an accident it would cause harm to people in Shizuoka. But worse still, if any fallout went

in the direction of Kanagawa and Tokyo, as it probably would during some months of the year, the consequences would be unimaginable. Indeed, we argued, as we still do, that Hamaoka is the most dangerous nuclear plant in Japan.[17]

Other groups comprised a broad spectrum of local residents, including lawyers, shopkeepers, school teachers, housewives, and some local farmers. Ono Yasuhiro, a prominent lawyer and historian, compiled a radical interpretation (at least in the eyes of the ruling elite) of Hamaoka's history. The town rejected it as an official history and refused to publish it.[18] Ono later became an important figure in the alliance with fishing cooperatives. These citizens' groups, like Leftist parties, were uncertain about the risks of living close to a nuclear power plant.

The Hainan fishing cooperative operated to the east of Hamaoka. As shown in map 3, it comprised the Omaezaki, Sakai hirata, Sagara, Jittogata, and Yoshida fishing cooperatives. This cooperative, while smaller than the Nanto cooperative, was the largest in Shizuoka. Its membership consisted a high proportion of deep-sea fishermen. Seventy percent of the catch was made up of deep-sea tuna, and the remainder was of coastal fish, predominantly whitebait. The coastline from Hamaoka to Yoshida was endowed with a shallow shore reef that provided an ideal environment for whitebait. Coastal fishermen developed special drag nets to increase production.[19] In contrast, almost all of the income of the Nanto fishing cooperative was derived from coastal pearl production.

Coastal fishermen were particularly worried about the environmental impacts of the project. The major concern was the possible reduction in the value of their catch. They argued that waste water discharged from the plant would increase the temperature of the water around the shore reef, the breeding ground for whitebait. Like fishermen from Ashihama, many actually thought the waste water would boil the ocean. Fishermen argued that even small temperature or radiation increases would adversely affect whitebait breeding and be detrimental to the longer term viability of the industry.

These fishermen found a leader in Hata Toju of the Sagara fishing cooperative. Sagara was more dependent on whitebait trawling off Hamaoka than any of the other cooperatives and stood to lose the most. Hata joined the JCP in 1955 and was also ideologically opposed to nuclear power. He became chairman of the Ogase JCP in 1958 and head of the Sagara fishing cooperative in 1963.[20] Hata, in his capacity as both a fisherman and a Communist, played a pivotal role in mobilizing resistance by fishermen.

In July 1967, Hata led a small group of fishermen to Ashihama on a study tour. He concluded that the underlying factor in the success of the Nanto fishing cooperatives was the formation of an alliance that allowed them to

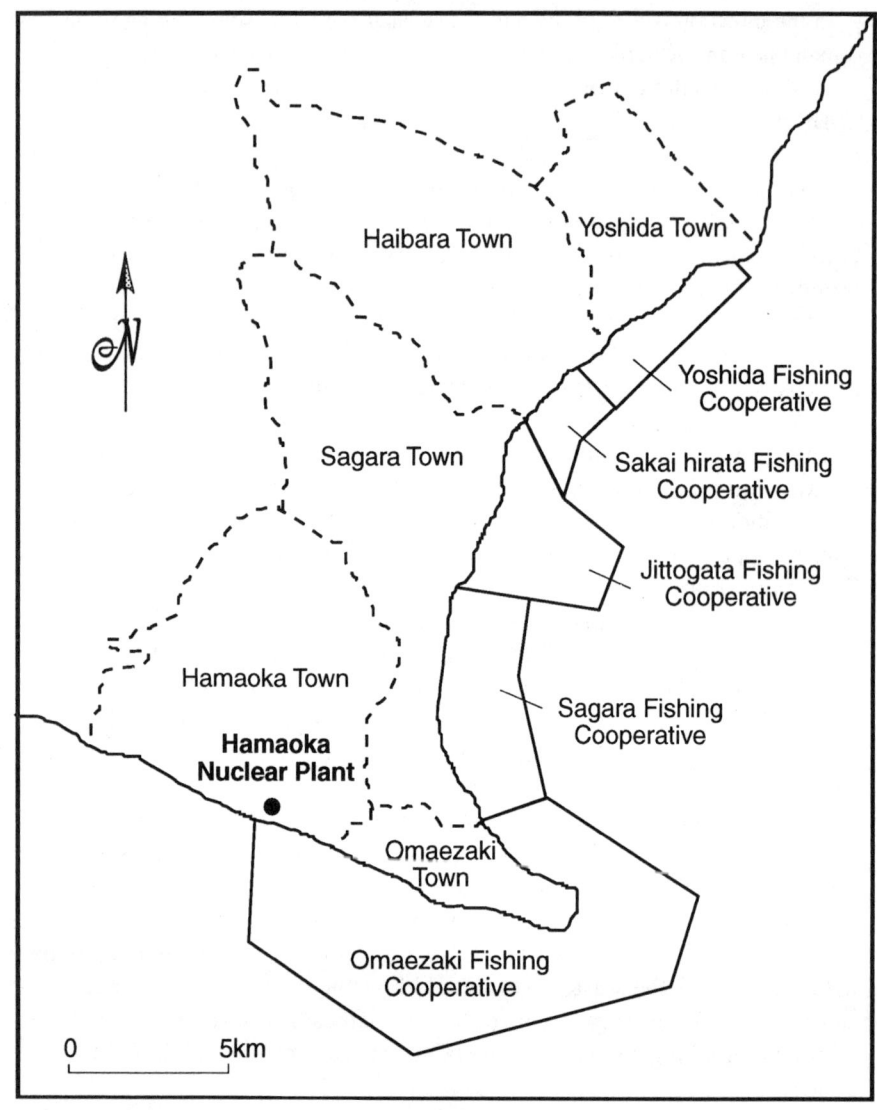

Map 3 Hamaoka nuclear plant: site location and property rights ownership

pool resources and increase their strength in the local assembly. Upon returning, he established and was elected the leader of the Alliance for Opposing the Development of the Hamaoka Nuclear Power Plant (Fishing Alliance), which included all of the cooperatives in Hainan. The Fishing Alliance adopted a policy of absolute opposition to the nuclear plant.[21] Seeking to emulate Nanto fishermen, Hata's strategy was to get into power in Sagara town.

In September 1967, Chubu Electric, in conjunction with the Hamaoka administration, declared its intention to start negotiations with land owners. In response, the Fishing Alliance and the other anti-nuclear groups decided to join forces to oppose Chubu Electric and the town. In early October, Hata and Ono, the leaders of the main opposition groups, formed the Antinuclear Struggle Committee (Struggle Committee). The views of Hata and Ono were similar. Both opposed nuclear energy on safety grounds, regarding it as an experimental technology not ready for commercial introduction. They also opposed it on ideological grounds, seeing it as a technology that further cemented monopoly capitalism. Although many fishermen were not compatible ideologically with Ono's views, they were concerned about the situation. They believed that the utility and the Hamaoka administration were developing the project at their expense and were willing to downplay ideological differences for an improved bargaining position with the utility.

Hata and Ono saw the alliance as the most efficient way of increasing their political capacity to oppose the project. They assessed that each group had weaknesses that would inhibit the group's ability to oppose successfully if they acted unilaterally. The movement, consisting of citizen and Leftist elements, was relatively small and did not possess a strong organizational support base in Hamaoka or at the prefectural level. The Hamaoka assembly was dominated by the LDP. There was only one JCP assemblyman, and the JSP had no representation. The movement was financially weak and geographically dispersed. Although the citizens' groups were based in Hamaoka, Leftist political groups were mainly from outside of the town. Hamaoka did not have a history of labor movements, rural uprisings, or collectively repelling outsiders.[22]

Unlike the Ashihama case, the local opposition at Hamaoka had little political, moral, and financial support from political parties at the prefectural level. Although the JSP and JCP jointly held approximately 40 percent of the prefectural assembly seats, they had lost five seats in the 1966 election and were uncertain about the electoral implications of opposing a plant that was portrayed as being in the community's economic interest.[23] Furthermore, Leftist political parties were not overly concerned with environmental degradation in Hamaoka, as electoral reliance on that town in prefectural elections was extremely low. The movement had only limited access to decision making at the local and prefectural levels, which prevented it from mounting effective opposition.

Hata found it difficult to get strong support from the majority of fishermen. Only a quarter of fishermen in the Hainan fishing cooperative were involved in a demonstration at Sagara in August 1967. This reflected not only their ideological differences but also the relative economic size, prospects, and structure of the cooperative. The cooperative employed only 30 percent of the regional population. In contrast, the Nanto cooperative was larger, employing 70 percent of the regional population. Prospects for the two cooperatives also differed. The annual value of fish caught by the Hainan cooperative totaled 300 million yen, and there were few prospects for expansion. The annual value of the pearls in Nanto was 3,000 million yen, and the industry was experiencing a pearl boom with strong growth expectations.

The potential impact of waste water from the plant also differed between the cooperatives. In Nanto, given production specialization, the expected adverse effects were evenly spread across the industry. In contrast, the potential effects the project differed markedly within the Hainan industry. Whereas waste water was expected to affect coastal fishermen adversely, it was not perceived to have a negative impact on deep-sea fishermen who used the Omaezaki port only as a base for their distant operations. Moreover, many deep-sea fishermen were interested in obtaining compensation to help alleviate their debt problems without having to relinquish property rights.[24]

To have a chance of success, Hata and Ono knew they needed to position themselves quickly. They were aware that the utility's bargaining position would strengthen greatly with completion of land negotiations in Hamaoka. It would own the land and could with the permission of the town commence detailed investigations and preliminary construction. As there were no fishing cooperatives based in Hamaoka, Hata and Ono presumed the town would hesitate only briefly in issuing permits. The movement targeted land negotiations by sensitizing the safety issue. They hoped that added uncertainty about nuclear risks would dissuade some land owners from selling their land. They distributed pamphlets, conducted car parades and boat demonstrations, collected signatures from local residents, and appealed to local and prefectural assemblies.[25]

Opposition to unwanted projects seeks to alter the structure of bargaining relationships through protest and coalition building. The bases of opposition differed fundamentally between Hamaoka and Ashihama. The first was that the expected negative impacts were spread more unevenly across fishing cooperatives in Hainan (Hamaoka) than in Nanto (Ashihama). This unequal distribution weakened the ability of the former to engage in collective action. The second difference was that the fishing industry was less important to the prefectural economy in the Shizuoka compared with the Mie prefectural economy. Resistance at the former had much less outside financial and political support for its position. The third difference was that resistance at Hamaoka relied on joining forces with ideological resistance to strengthen its bargaining position whereas resistance at Ashihama did not. While the resistance at Hamaoka was much weaker than that at Ashihama, it still had the potential to destabilize and block negotiations.

RISK AND DISTRIBUTIONAL CONCERNS

Chubu Electric knew about the Hainan fishing cooperative to the east of Hamaoka. Indeed, one of the reasons it found the Hamaoka site so attractive was that there were no fishing cooperatives in the town. The company was, however, unaware that the Hainan cooperative had inherited the fishing rights of the then-defunct Sakura fishing cooperative from Hamaoka in 1962. This ownership came about through the amalgamation of the Sakura and neighboring Omaezaki fishing cooperatives. In 1963, other nearby cooperatives united to form the Hainan fishing cooperative. The cooperative had a legal and legitimate claim to compensation.[26] A company official summed up the reaction to the emergence of fishing opposition.

> Initially we thought that resistance was simply designed to extract money from the company. Our early investigations suggested that the Hainan cooperative did not have property rights. We were stunned to learn that it actually did and that we would have to negotiate with it to site the project. We obviously did not complete these investigations thoroughly enough. We knew from Ashihama that fishing cooperatives could create havoc in negotiating processes. When our president found out about this, he was visibly angry. We all thought that it would be another Ashihama, all over again. We knew that heads would roll if that turned out to be the case.[27]

Company officials were fortunate the size and structure of the fishing industry in Shizuoka differed from that in Mie.

The anti-nuclear movement heightened awareness within the community about the potential environmental hazards of nuclear energy. This public cognizance came at a time when environmental concerns were becoming more pronounced throughout Japan, although still not as strong as during the early 1970s. The utility devised a strategy to alleviate safety concerns. Together with the town, it held a number of public lectures and visited local residents to discuss their worries. These activities not only provided information that downplayed nuclear risks but also allowed promoters to assess the varying response patterns of community interests toward the project.

Promoters persuaded different sections of the local community to visit other sites, such as Tokaimura and Mihama.[28] The Tokaimura plant had a good safety record, and Mihama was experiencing a construction boom. Chubu Electric financed the visits. Local community groups from Hamaoka discussed safety and regional development issues with local groups and utility personnel at these sites. Promoters carefully sequenced the visits from the top levels of society down: first, town politicians and officials; then, heads of neighborhood associations and hamlets; and then other regional groups, such as housewives and school children.

This approach changed community perceptions about the risk and benefits of nuclear power. It highlighted the significant economic rewards the town would reap from hosting the nuclear power plant. It also demonstrated that other regions had accepted nuclear risks in return for these observable benefits. Moreover, the company did not have to deal with an articulate environmental movement at that time. Social attitudes generally placed more emphasis on the expansion of social and economic opportunities than on environmental degradation. The visits were extremely successful in reducing community concern about nuclear safety and weakening the legitimacy of the anti-nuclear movement's message. As one resident remarked,

> The visits really changed our attitudes about nuclear safety. We talked with many other residents at Tokaimura. They said that if they were not afraid of nuclear power, why should we be. One picture (or in this case, one trip) was really worth more than a thousand words. We also had a great time. We did not pay anything, and whoever was footing the bill provided a lot of entertainment. They gave us a lot to drink (sometimes we drank too much), not to mention the feasts they put on and even *omiyage* [souvenirs] they bought for our children.[29]

Chubu Electric clearly managed safety concerns very effectively. It used the demonstration effect in a structured way to reinforce, from the top to the bottom levels of the community, the attitude that nuclear power offered big benefits and posed little risk. In contrast, the anti-nuclear movement was not endowed with the finances to take local groups to other sites, such as Tsuruga, which had experienced nuclear accidents. The movement's approach of relying only on lectures, demonstrations, and the distribution of pamphlets reflected its limited resource base and its lack of support at the town and prefectural levels.

Having weakened its opponents, the utility proceeded to make its first offer to land owners through the town on 5 February 1968. The offer led to unrest among non-owners of property rights in Sakura hamlet. They argued that they would incur equal risks but would not receive direct benefits, such as side payments, or even any guarantees of getting indirect benefits, such as employment opportunities.[30] In short, though they were reasonably certain about the costs, they were very uncertain about the benefits. Such intra-local (as well as other intra-regional) distributional issues were a major factor in the establishment of the Three Laws in 1974.

The town responded quickly by creating the Sakura Hamlet Policy Committee on the 11 February 1968. This committee, chaired by Kamogawa, proposed that the utility provide community compensation to the hamlet. This proposal led to a formal agreement between the committee and the utility covering community compensation for infrastructure development and guaranteed employment opportunities for hamlet residents. This strategy provided direct and indirect benefits to non-owners of property rights and reduced their resistance to the project. The utility had clearly learned that failing to address intra-community distributional concerns could have dire consequences for siting, even when local political leaders were strongly committed to projects.

Despite the market need for the project, the power company wished to delay actual payment of community compensation until land negotiations were completed. It argued that the immediate payment might improve the bargaining position of land owners, who might tactically increase their demands for relinquishing their land rights. The utility wanted to schedule community compensation and property rights transfer payments at roughly the same time. The town agreed to provide a guarantee to the Sakura hamlet that compensation would be paid. As an interim measure, it borrowed financing from the prefecture to construct a cultural center that was completed in late August 1968. This gesture showed the community that actual, as opposed to potential, benefits resulted from accepting the project.[31] The cultural center symbolized that offers of compensation would be met and that agreements would not be abrogated.

Chubu Electric made a second offer to land owners through the town in early March 1968. This proposal was again rejected by some land owners. There remained a difference between what the utility was prepared to pay and what land owners were prepared to accept for transferring their land rights.[32] The utility had based its first offer on the "income derived payment formula" contained in the Compensation Standards, and increased that marginally in the second offer.[33] Landowners based their demands on the "similar situation derived payment formula" also contained in the Compensation Standards. They demanded roughly the same amount paid in the settlement between the prefecture and land owners over the construction of the Tokai Expressway. Landowners in that case had received more than Chubu Electric was prepared to pay to Hamaoka landowners.

Although these differences reflected the nature of property rights transfer arrangements, they also partly reflected different interests of land owners. The cost of selling property rights was not spread evenly among different groups of land owners, and the motivations of these groups differed substantially. The largest group, consisting of part-time farmers whose land was unproductive, was willing to release property rights at a relatively low price. Others were less willing to accept the utility's offer. Specialist full-time farmers generally opposed parting with their rights. Others delayed to bargain for higher transfer payments. Some individuals were just uncertain as to whether or not they should sell.[34]

The lack of homogeneity weakened the groups' collective bargaining position. Yet, at the same time, it provided an opportunity for the Leftist antinuclear movement, which included some farmers, to stifle land negotiations. Chubu Electric and the town became very concerned that resistance might emerge in Sakura hamlet if guaranteed community compensation were overly delayed.[35] It became obvious that siting the project hinged on the rapid completion of land negotiations.

The town leadership responded by establishing the Land Negotiation Committee in early May 1968. The committee head was Kurebayashi Matsutaro, deputy director of the Sakura agricultural cooperative and a close friend of Kamogawa. The committee played a crucial role in the land negotiations. First, it isolated the majority of property rights owners from the Struggle Committee and allowed landowners to rely on the Land Negotiation Committee to resolve conflict over the economic valuation of their property. Second, the committee allowed Kamogawa and Kurebayashi to influence landowners by stressing the social decision-making rules of the hamlet, which placed an emphasis on working together for the good of the hamlet as determined by the leadership. Third, the committee created a unified partner with which promoters could bargain, relatively free from external interference.[36] Negotiations were completed rather quickly once they commenced.

Even relatively weak opposition movements can often influence siting by sensitizing risk issues and raising community uncertainty about gains from accepting projects. Promoters at Hamaoka responded to opposition strategies very effectively. They used the demonstration effect in a structured way to maximize the political effects of showing that others were willing to accept risks in return for project benefits. This strategy reduced community uncertainties about risk. Promoters used symbolic compensation to manage intra-community distributional concerns and showed that the company would live up to its promises. The ability of the utility to change community risk perceptions was crucial in land negotiations. It allowed promoters to isolate property rights holders from ideological interests and negotiate a deal.

THIRD-PARTY MEDIATION

The emergence of the Struggle Committee caused the prefectural government to worry about possible community reactions about radiation. In 1954, Kubokawa Aiikichi, a telecommunications officer from Yaezu city, died as a result of radiation exposure from the Bikini nuclear test.[37] This historical experience and the location of the plant near a fault line strengthened the potential for community concern about the risk of radiation relative to other nuclear regions, such as Fukushima and Mihama. The antinuclear movement used Kubokawa's death to sensitize the prefectural community to nuclear risks in the hope of destabilizing fishing negotiations.

The prefecture wanted to weaken the Leftist movement but neither it nor Chubu Electric wanted to be perceived as meddling in local affairs. In 1967, the prefecture, together with large private interests, pressured local Numazu residents to accept an oil project. That effort resulted in immense local resistance, and the prefecture abandoned the project in the same year.[38] The prefectural leadership judged that similar tactics would probably engender local resistance to the project, providing additional stimulus to the Struggle Committee. The experiences at Ashihama and Numazu still remained fresh in the minds of Chubu Electric's negotiators and prefectural leaders.

Chubu Electric would not commence negotiations with fishing cooperatives because the Leftist movement could block them through Hata's position as leader of the Fishing Alliance. The prefecture responded through backdoor political channels. In late December 1968, the prefectural branch of the LDP established the Special Nuclear Energy Committee (Special Committee). Its aim, at least publicly, was to consider the nuclear safety issue.[39] But it also had another agenda, which was to split the Struggle Committee by breaking the nexus between the Fishing Alliance and the Leftist anti-nuclear movement. Given Hata's ideological position, it sought to remove him from the leadership of the Fishing Alliance.

The strategy consisted of several carefully engineered steps, which were portrayed as originating from the regional level. The first step was to establish the Omaezaki and Sagara Nuclear Committee (Nuclear Committee), which the Special Committee did on 20 January 1969.[40] It comprised two groups: local residents from Omaezaki and Sagara towns and members of the five Hainan fishing cooperatives. Both groups studied nuclear safety and regional development. The Nuclear Committee stressed that regional affairs should be conducted independently, without outside influence. This approach received overwhelming community support.[41] Local autonomy can work against outside ideological interests just as much as it can against outside private, bureaucratic, and political interests.

There was great concern among regional and prefectural promoters about the possibility of opposition in Sagara and Omaezaki. Chubu Electric planned to use existing port facilities at Omaezaki to construct the plant and for fuel loading, because developing new facilities at Hamaoka would have increased project construction costs considerably. The economic merits of the project to both towns were markedly less than those to Hamaoka.

Both could not expect large economic flow-ons from construction. Furthermore, the risks associated with plant were not limited to Hamaoka's administrative boundaries. There was concern in surrounding areas about environmental effects, particularly on mandarins and *sake*, important mainstays of their local economies. The Nuclear Committee provided a forum for discussing nuclear safety and regional development issues to counter attempts by the anti-nuclear alliance to recruit more support from Sagara and Omaezaki.[42]

The second step was to promote a change of attitudes within the Fishing Alliance. Yanagihara Seiji, a prefectural politician in the Special Committee, played the lead role. He persuaded Kawaguchi Yuzo, head of the Omaezaki fishing cooperative and president of Kawaguchi Tekko, a steel company in Shimizu city, and Haraguchi Inaichi, head of the Jittogata fishing cooperative, to oust Hata. Yanagihara owned the Shin Yanagihara Boat Factory in Yoshida, a city neighboring Sagara to the north. He provided financial assistance to Kawaguchi's steel factory and political support to Haraguchi in fishing cooperative elections.[43]

Kawaguchi and Haraguchi appealed to the deep-sea fishermen to alter the attitude of the Fishing Alliance. These fishermen had earlier supported coastal fishermen, but Kawaguchi and Haraguchi persuaded them that they would gain from compensation, which would help pay off their debts. Deep-sea fishermen were receiving prefectural subsidies, mainly due to Yanagihara's influence at the prefectural level, to expand port facilities at Omaezaki. Yanagihara, through Kawaguchi and Haraguchi, argued that continuation of these subsidies might be jeopardized if deep-sea fishermen did not support the project. Deep-sea fishermen were numerically dominant in the Fishing Alliance, and this change in attitude altered the position of the alliance as a whole.[44]

On 27 March 1969, Kawaguchi and Haraguchi set up the Board of Director's Nuclear Study Deliberative Committee (Director's Committee), which consisted of the heads of the five Hainan fishing cooperatives. The Fishing Alliance stated that it would not oppose the project if it were safe and if an investigation of the impact of the project on the regional fishing industry were conducted in conjunction with the Nuclear Committee.[45] The Director's Committee commissioned a group, consisting of prefectural officials and fishermen, that spent two months investigating the major concerns of coastal fishermen. The committee convinced Chubu Electric to extend the length of the waste water pipes so that they would not discharge waste water close to the whitebait breeding areas. Risk mitigation weakened the arguments of coastal fishermen, isolated Hata, and weakened his joint opposition with Leftist elements.

The third and final step was to oust Hata as leader of the Fishing Alliance, install a new leader, and finally bring the Fishing Alliance into the Nuclear Committee. Haraguchi persuaded Nakamoto Ichiro to change Hata's stance. Nakamoto was one of the directors of the cooperative and had extremely close personal links with Hata. He had supported Hata in fishing cooperative elections and persuaded Hata to participate in the fishing investigation.[46] Pressure on Hata from within the Fishing Alliance and within the Sagara cooperative effectively isolated him from the mainstream position. On 15 April 1969, Onada Shosaku, a cousin of Kawaguchi, defeated Hata in an election for head of the Fishing Alliance. Ousting Hata effectively broke the nexus between the Fishing Alliance and the Leftist anti-nuclear movement. As one close observer mentioned, "The political pressure to remove Hata surgically from the leadership of the Fishing Alliance was immense. It reminded me of what one hears about CIA operations. I do not fully understand why he changed his position about the investigation. Some say he was bought off. I do not know if this is true or not. I do not want to talk about it, as he has been in a psychiatric ward in a local hospital ever since."[47]

The Director's Committee not only investigated safety concerns; it also started fishing rights transfer and compensation negotiations with Chubu Electric. After ousting Hata, it changed its name to the United Fishermen's Council on Nuclear Energy (Fishermen's Council). Despite continued resistance by some coastal fishermen (mainly Hata's followers), it commenced preliminary negotiations with the utility. By this time, Chubu Electric had reached broad agreement with the Fishermen's Council over the transfer of fishing rights. As it also had agreement from relevant local mayors, the prefectural governor gave his approval for the project.

The company had completed the requirements for submission of the proposal to the EPDCC. By this stage, the expected electricity supply situation had worsened. Chubu Electric appealed to MITI to support the project.[48] The national government was receptive because the supply situation in the central electricity sphere also had worsened. In May 1969, it granted conditional EPDCC approval subject to the satisfactory completion of fishing rights negotiations. This conditional approval allowed the utility to start obtaining licenses and preliminary construction while it negotiated final fishing rights transfer arrangements.[49]

The ability of third parties to mediate is crucial in bargaining over unwanted facilities. As at Ashihama, mediators were brought in to resolve the dispute between the utility and property rights holders. In contrast to Ashihama, where pressure was exerted directly and overtly from the top, however, promoters at Hamaoka used local, behind-the-scenes power brokers to manage the dispute. They changed the expectations of local fishing groups toward the project by getting the utility to make minor design changes. These changed expectations enabled them to oust the leader of the opposition alliance and install a new one who was more willing to negotiate. The leadership change altered the bargaining environment and allowed the company to commence negotiations. Even though mediators exerted enormous political pressure on the opposition, the leadership maintained its legitimacy because it used local power brokers and not higher level governments.

CONCLUSIONS

Promoters used very effectively a range of compensation instruments to manage opposition toward the Hamaoka project. These included compensation for land and fishing rights (600 million yen) transfer, community compensation (1,260 million yen), risk mitigation through the use of the demonstration effect and minor design changes, sociopolitical and symbolic compensation through status on regional committees and social infrastructure, and some reported bribery. In contrast with the Ashihama and other disputes examined in this book, the utility obtained much faster agreement with significantly less compensation.

The structure of the bargaining environment was shaped in a way that, according to the polimetric results, would have resulted in a much shorter settlement time (fifteen months) compared with the average of initial nuclear sitings in Japan (eighty-two months) and certainly much shorter than Ashihama. The utility and the national government faced large electricity shortages in their respective electricity spheres. Income growth was relatively high in the prefecture, and there was a strong community desire for further expansion. The financial state of local government was relatively unhealthy, and the government saw the project as important in that it would provide more local infrastructure. Economic opportunities in the rural sector were very low, and rural interests placed little emphasis on the environmental costs of the project. Local Leftist political party representation was comparatively weak. Social attitudes attached relatively little concern on environmental degradation.

These results shed light on the overall structure of the bargaining environment in which participants resolved the conflict. The utility and the national government placed a high market need for the project and wanted a quick settlement. Subnational governments promoted the project to achieve local and prefectural economic development objectives. Other community interests, including rural ones, did not place strong emphasis on the environmental and other costs of the project. Collectively, these conclusions suggest an environment where promoters had more than sufficient compensation to manage community resistance.

In contrast to the Ashihama case and others examined, the statistical results were much more consistent with actual siting times, which is reflected in the small residual (eight months). Nonetheless, we need to understand how promoters dealt with an opposition that enhanced its power by forming a coalition. The utility managed this opposition effectively because it learned from the Ashihama dispute that weak opposition can readily become intense. It managed risk and intra-community distributional concerns and used behind-the-scenes power brokers to split the alliance. This strategy prevented opposition forces from altering the structure of political power and allowed promoters to negotiate a deal with key stakeholders.

The expected costs of the plant were, in contrast to the Ashihama site, spread unevenly across local and regional rural interests. This distribution constrained their ability to oppose the project. For example, although coastal fishermen expected to be adversely affected, initial opposition by deep-sea fishermen was turned into support for the project. The structure of the fishing industry was not characterized by specialization in production. The per-capita benefits of engaging in collective action differed among the two groups, which prevented the industry from sustaining a united front in its resistance. Many deep-sea fishermen did not participate in opposition demonstrations. The lack of strong collective actions weakened the industry's bargaining position.

The allocation of bargaining power in the dispute consistently penalized opposing interests over supporters of the project. The utility wanted to develop the project, and the Hamaoka community also generally wanted it. Unlike the opposition from Nanto at Ashihama, community interests in Hamaoka had fewer viable alternatives to negotiating an agreement with the utility. Even though the opposition increased negotiating costs to the utility, this opposition did not reach a level that made the power company prefer an alternative location. The opposition's weak bargaining position reflected the structure of the resistance and its inability to gain any access to local political power.

Project promoters, in contrast to the Ashihama case, managed the opposition effectively. The resistance formed an alliance to enhance its bargaining power and targeted its strategies to destabilizing crucial rights transfer negotiations by sensitizing uncertainties about nuclear risks. Promoters countered these strategies systematically throughout the course of the conflict. Local mediation was crucial in maintaining political legitimacy. Local authorities managed risk and intra-community distributional concerns through risk mitigation and community compensation strategies. Local power brokers were instrumental at all stages of the siting process. They delayed initial siting decisions to avoid intra-party competition influencing bargaining processes. When the conflict intensified, power brokers isolated property rights holders from ideological opposition to minimize extraneous interferences influencing negotiations.

Expectations about the structure of the bargaining environment changed during the conflict. In contrast to the Ashihama dispute these expectations changed in ways that brought the conflicting parties together. Endogenously induced events changed expectations. The unexpected emergence of the opposition coalition shocked the utility. Local and regional political leaders split this resistance and created positive expectations about resolving the dispute. Managing safety concerns and delivering symbolic community compensation at a time when local residents were questioning the value of the project prevented community dissatisfaction in Hamaoka from turning into powerful opposition. Negotiators changed expectations in ways that increased the gains and reduced the costs to key stakeholders of accepting the project.

As at Ashihama, conflicting parties faced high levels of uncertainty. Local and regional interests were uncertain about the risks of nuclear power. The rapid emergence of opposition heightened uncertainty for the utility. The structured use of the demonstration effect reduced community concerns about nuclear risks. Changing the position of the waste water outlet pipes helped alleviate concerns among coastal fishermen. During the course of the dispute, promoters used innovative strategies to reduce uncertainty to a level that enabled them to strike a deal with community groupings.

Bargaining and compensation is a learning process. Although the utility still misjudged fishing opposition (a fundamental mistake), it recovered from this far more effectively than it had Ashihama. First, it used pressure and influence from local power brokers, rather than from the top, to mediate in the dispute. Second, it intentionally delayed some decisions until intraparty conflict was managed. Third, it employed legitimate strategies to keep opposition from getting direct access to political power. Fourth, it addressed intra- and inter-community distributional concerns carefully to prevent escalation of the siting conflict. What mattered most was that it learned not to attempt to bypass local and regional opposition but rather to negotiate with that opposition.

Hamaoka shows that NIMBY disputes do not always take long times to resolve or end up in permanent gridlock. Even though the parties started with widely and fundamentally different positions, the use of bargaining and compensation resolved the siting conflict quickly. Promoters carved up and managed opposition in creative ways. As we will see in chapter 6, Hamaoka is not an isolated case of obtaining relatively quick community agreement to noxious projects. But, surprisingly, in the Matsushima case the core resistance came less from local and regional community interests than from private and national bureaucratic interests.

C · H · A · P · T · E · R   S · I · X

Capitalizing on External Shocks

Resolving siting disputes is a dynamic process. Expectations of at least one participant will need to change if conflict is to be resolved. Changes in expectations become important when they are large enough to alter perceptions of critical stakeholders. Revised expectations can feed back into negotiating processes by affecting the structure of bargaining power and bargaining relationships. The extent to which promoters and developers capitalize effectively on such changes and employ compensation strategies is decisive in determining whether siting conflicts can be resolved.

In just under four years, the Electric Power Development Company's (EPDC) received approval for its Matsushima coal-fired plant. This time frame was slightly longer than the average of about three years for all initial fossil-fueled projects and significantly longer than many of the more easily sited nuclear plants, such as Hamaoka. The Matsushima case is atypical of initial siting disputes. Resistance from extra-local market and bureaucratic interests, not the community, initially stalled bargaining and resulted in a longer settlement time. Matsushima shows how intermarket and interjurisdictional conflict, outside the confines of local communities, can delay project siting.

It is intriguing that the utility had to pay very large amounts of compensation to win approval for the project. Changed expectations allowed promoters to break an impasse and negotiate a settlement with opposing market and bureaucratic interests. This settlement required the EPDC to compensate a private power company. Yet, in managing this extra-local conflict, the company broke a promise it had made to the local community and lost its credibility with some important community interests. Even though the EPDC had Ministry of International Trade and Industry (MITI) backing and there was virtually no organized resistance, it still paid more compensation to fishing and community interests than in many nuclear siting disputes. In bargaining, it is crucial to maintain credibility even if one's bargaining position improves suddenly and dramatically. Failure to do so can strengthen the positions of local interests.

REGIONAL WELFARE POLICY AND PRIVATE OPPOSITION

The EPDC was established in the early 1950s as a MITI-affiliated company. Seventy percent of its financing is provided by MITI; the remaining funds come from the nine electric utilities.[1] Unlike private power companies, the EPDC does not supply retail electricity directly to consumers. Rather, it provides wholesale electricity to private utilities. Officially, the EPDC acts as a public insurance company, reducing risks of electricity shortages. Until the late 1950s, it mainly developed hydroelectric power, but, as hydroelectric sites became exhausted, it started to build coal-fired plants. This shift in emphasis also reflected its policy role in supporting the domestic coal industry.[2]

Coal was the major form of energy used in Japan in the immediate postwar period. In the mid-1950s, large oil discoveries in the Middle East reduced the relative price of oil. Although there was strong coal industry opposition to interfuel substitution, the government used financial mechanisms, mainly through subsidies and taxes, to expand the use of oil to stimulate Japanese economic growth. In response to these market and policy incentives, the power industry moved rapidly from using coal to oil.[3]

Production declines created economic problems in coal-producing areas. Bipartisan pressure from Liberal Democratic Party (LDP) and Japan Socialist Party (JSP) politicians, who had support bases in those areas, forced MITI to formulate policies that subsidized the use of domestic coal and restrained its declining use. Between 1953 and 1972, MITI's production targets were never met. As a result, MITI set consumption targets in 1972. Although it persuaded power and other industries to purchase coal, the EPDC became the largest single domestic coal consumer in the power industry.[4] It also became the largest single welfare provider to Hokkaido and Kyushu, the two major coal-producing regions.[5]

Unlike the private utilities, the EPDC does not possess a specific regional electricity sphere. In principle, the EPDC can site power plants anywhere in

Japan. Technical site selection criteria and siting processes are similar to those that apply to private power companies but with two important qualifications. First, as the EPDC builds projects in private utility spheres, it must also obtain consent from relevant power companies. Second, as it receives funds from the national treasury, the Ministry of Finance (MOF) must approve the financing of its projects.[6] As a company official remarked, "Siting is tough enough in Japan. But for the EPDC it is even tougher. We have to negotiate not only with local and regional interests but also with other utilities and the national bureaucracy, all of which have veto power over our siting decisions. This is odd given that we are a public insurance company and site projects that the private utilities cannot or will not develop."[7]

The EPDC focused on Hokkaido and Kyushu as potential sites for a coal-fired plant. It wanted to minimize transport costs by locating a project as close as possible to coal-producing areas.[8] Developing a fossil-fueled project in Hokkaido would not be easy. During the early 1970s, MITI and the Hokkaido government were already putting considerable pressure on Hokkaido Electric to develop the Tokoma-Atsuma coal-fired plant to increase the demand for domestic coal in Hokkaido.[9] With that project and the Date oil-fired plant, Hokkaido Electric expected to be able to meet electricity demand increases (see Chapter 7).

The EPDC subsequently focused its attention on Kyushu and started scanning for sites in early 1973. Investigations revealed a suitable site in a small hamlet called Uchiura on Matsushima island off the coast of Ooseto town in Nagasaki prefecture, approximately midway between Nagasaki and Sasebo, the two main consumption centers in Kyushu, relatively close to Kyushu Electric's transmission network. As shown in map 4, the island, which is sparsely populated, lies within the administrative jurisdiction of Ooseto. A deep port would allow access for vessels transporting coal. Matsushima Kosan, a company that had developed coal mines on a number of Nagasaki's offshore islands, owned 70 percent of the land.

The EPDC set out to create regional support for the project. In February 1973, after discussions with EPDC officials, the Nagasaki Regional Liaison Committee for the Coal Industry (Liaison Committee) expressed strong interest in the project. The committee consisted of local mayors from coal-producing areas and was a powerful lobby group in Kyushu. It sent petitions to a number of local governments, including Ooseto, requesting that they consider hosting a project. In April of that year, it also requested that MITI cooperate.[10]

The prefectural government viewed these developments positively. The attainment of regional social and economic objectives dominated the environmental concerns. In June 1973, Governor Kubo requested financing for the project from MITI and MOF. He gave five reasons to justify his request:

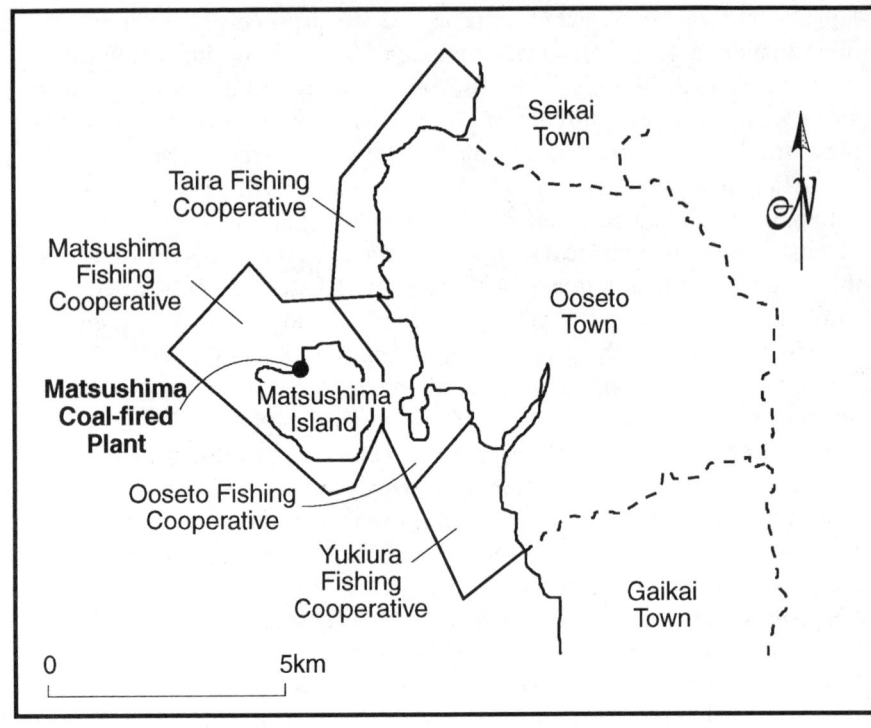

Map 4 Matsushima coal-fired plant: site location and property rights ownership

the need to mitigate devastating economic impacts of mine closures, the need to guarantee a stable demand for domestic coal for social policy reasons, the need to develop coal-fired plants for the long-term viability of the coal industry, the need to maintain industry employment levels, and the need to stabilize the industry to contribute to regional economic development.[11] There was no mention of electricity markets in Nagasaki or in the broader Kyushu electricity sphere. Kubo's goal was regional welfare, not an electricity market objective.

Despite strong support from the local and prefectural political elite in Nagasaki prefecture, Kyushu Electric opposed the siting.[12] As far as it was concerned, "The idea of locating a coal-fired power plant in Kyushu at that time was ludicrous. There was definitely no market justification for the project, and we perceived it as being very costly to us. It was crazy for the EPDC to even think about putting an additional coal-fired plant in Kyushu."[13] Of course, Kubo did not include a market justification for the project.

Kyushu Electric argued that there was no economic rationale for additional capacity in either Nagasaki or Kyushu and that Nagasaki was a net exporter of electricity.[14] In 1973, the utility supplied 5,716 megawatts (mw), and five-year projections indicated that demand would increase to 8,210 mw. The company expected to be in a favorable supply position as a number of its plants had substantial capacity (2,974 mw) and were either in the licensing stage or under construction.[15] Because the Kyushu grid was interconnected with the Honshu grid by only a small transmission line, Kyushu Electric argued that it could not distribute much excess capacity to the mainland. It did not wish to purchase costly electricity that it could generate more cheaply.

Interestingly, the company did not accept the security argument. Coal mines had been closing rapidly in Kyushu. The industry had phased out coal much more quickly than government plans anticipated. As a result, Kyushu Electric experienced domestic supply problems and was forced to modify several coal-fired plants to burn both oil and coal. The cost of these design changes was high compared with the cost of developing new oil-fired plants, even after factoring in relative fuel prices.

Kyushu Electric was also worried about the impact of the Matsushima project on electricity tariffs. The price of domestic coal had increased even with government subsidies and improved productivity. Mining companies had to mine deeper and more costly underground coal bodies, which was a major contributor to increased electricity prices in Kyushu. By the early 1970s, Kyushu electricity tariffs were the highest in Japan, causing a political problem for the utility. It did not want further upward pressure on electricity prices.

Kyushu Electric's resistance was also related to the project's potential impact on the environmental movement. Kyushu coal had a high sulfur content. The company argued that the project might lead to the emergence of strong local environmental resistance because Ooseto town had a relatively high population density. Anti-pollution movements had been emerging in Kyushu since the early 1970s and might intensify. The utility believed that a strengthened environmental movement would increase the difficulty of siting other projects elsewhere in Kyushu.

As far as Kyushu Electric was concerned, the project was going to lead to excess supply, which meant it would have to reschedule some of its projects. It argued that rescheduling would lead to political problems in areas where support had been created. Delay in delivering economic benefits to communities could result in a political backlash if expectations about income and employment expansion were not met. Moreover, many of the projects the company was developing were subsequent ones and community demand for them was strong (see Chapter 3). Kyushu Electric was also worried about higher interest costs on borrowed capital for nuclear projects, such as in Genkai, which were under construction.

The company's response to the EPDC project was also related to territoriality. In the late 1960s, larger utilities started to develop projects in smaller companies' electricity spheres. Site availability became limited because of rapid expansion and stronger urban environmental resistance. Tokyo Electric had developed the Fukushima nuclear project and was developing the Kashiwazaki-Kariwa nuclear project in Tohoku Electric's sphere. Kansai Electric had developed the Oi, Mihama, and Fukui nuclear projects in Hokuriku Electric's sphere. Kyushu Electric felt that larger utilities should not be developing scarce rural sites at the expense of smaller ones.

Kyushu Electric was in a strong bargaining position with the EPDC. The cost of the Matsushima project was high, and Kyushu Electric appealed to MOF not to support it. Furthermore, negotiations between the EPDC and Nagasaki community interests, on the one hand, and among the prefecture, MITI, and MOF, on the other, were being conducted secretly. Both the prefecture and the EPDC thought there might be resistance in Ooseto town if it were revealed that they had been conducting negotiations without the town's knowledge. The EPDC, like Chubu Electric at Hamaoka, decided to delay the siting process and not pressure Kyushu Electric into publicizing the dispute.

Noxious facilities can also impose negative spillover effects on other companies in the same industry, either indirectly as at Ashihama, or directly as in this case. These private interests are likely to oppose such projects even if they are vigorously promoting others. Kyushu Electric opposed the Matsushima project because the company was highly uncertain about the impacts of the project on its ability to maintain monopoly control over its electricity sphere. In contrast, the EPDC promoted the project to achieve political welfare policy objectives. The dispute highlighted divergent private and public interests over market control and political welfare objectives. Kyushu Electric's monopoly status in the regional electricity market gave it a strong bargaining position and forced the EPDC to delay siting. Private companies can cry NIMBY just as loudly as local communities.

REVISED EXPECTATIONS AND COMMUNITY RESPONSE

The 1973 oil crisis generated much more intense support for the project. MITI revised its energy policy and aimed to reduce dependence on imported oil. The EPDC's role expanded from the peripheral social one of providing an outlet for domestic coal to an important one as a major developer of coal-fired projects to reduce Japan's vulnerability to oil supply interruptions. Furthermore, in a period in which nuclear siting was experiencing increasing delays, coal-fired projects, which have shorter gestation times, were seen as the major bridging source of electricity until more nuclear projects could be developed.[16] In mid-March 1974, the EPDC announced a revised five-year plan including three coal-fired plants, one of which was the Matsushima project.[17]

The company promoted the project with renewed enthusiasm. In the same month, it announced that it wanted to start construction by mid-1975. As the national budget was passed annually in January, the company planned to negotiate simultaneously with the community interests, Kyushu Electric, and MOF. To meet its new construction goals, it had to obtain Electric Power Development Coordination Council (EPDCC) approval by late 1974, leaving about six months to obtain the necessary licences.[18] Less than a year was allocated for obtaining political approval, not a lot of time given the magnitude and complexity of the negotiating task.

Earlier in February 1974, the EPDC had applied to Nagasaki prefecture for permits necessary to conduct an Environmental Impact Assessment (EIA). Nagasaki prefecture, which was particularly hard hit by the oil crisis, granted approval and intensified its support for the EPDC project. The worldwide recession was causing major adjustment problems for the shipbuilding industry, the major pillar in the Nagasaki prefectural economy. The project presented an opportunity to manage the effects of the oil crisis on the regional economy.

At the same time, environmental concerns were becoming more pronounced in Japan, as evidenced by the growth of anti-pollution movements. The prefecture wanted to prevent disrupted project planning so it placed three conditions on the EIA. The first was that the company was required to publicize all results of the investigation, the second was the that project would be abandoned if results indicated any adverse environmental impacts, and the third was the that company would not start property rights or other community compensation negotiations until after the assessment was publicized. These conditions provided a political insurance cover for the prefecture, as it knew any emergence of environmental resistance would only enhance Kyushu Electric's bargaining power.

The EPDC promised publicly to abide by these conditions. In doing so, it separated the environmental investigation from the decision to construct the project. It believed the emergence of any community resistance would jeopardize concurrent negotiations planned with Kyushu Electric and MOF. As a company official lamented, "It was important to enhance and maintain the legitimacy of our siting approach not only in terms of dealing with local stakeholders but also in strengthening our fairly weak negotiating position with Kyushu Electric and MOF."[19] As both the Ashihama and Hamaoka cases showed, the extent to which utilities maintain legitimacy in managing environmental risk is decisive in bargaining and compensation processes.

The Ooseto town assembly agreed formally to the EIA in July 1974. In mid-1973, the EPDC approached Mayor Nagata about the project. Nagata sought to create community demands for regional development by proposing a paper pulp project. Such a project was less environmentally obtrusive, and Nagata believed he could create a consensus quickly. He planned to change the proposal to a coal-fired project after creating a positive development mood.[20] Nagata approached residents from Uchiura hamlet on Matsushima island and, in early January 1974, thirty-six residents sent a request urging the town to support the project. The assembly approved it unanimously and then formed the Special Committee for Inviting the Development of Industry (Special Committee) to proceed with the proposal.[21]

Subsequently, the Special Committee abandoned the paper pulp project and proposed a coal-fired plant, which as expected, was received with enthusiasm. There was little prospect for regional development in Matsushima. During the Taisho period, Matsushima was an important coal-producing area. The island economy was highly dependent on mining, with about 15,000 people employed in the industry. In 1934, floods forced the largest mine to close and by 1963 other smaller mines were exhausted.[22] As there was no other industry that could absorb the workforce, many young people left in search of employment opportunities. The community hoped the power plant would stimulate the island economy and encourage young people to return.[23]

Matsushima also lagged in the development of social and economic infrastructure. The outflow of young people left an aging population. There were no hospitals, and the only means of transportation to mainland hospitals was a dilapidated ferry that operated twice a week. On numerous occasions elderly people had died because they had not received medical treatment in time. In this discussions with local residents, Nagata stressed the range of infrastructure and welfare benefits that would accrue from hosting the project.[24]

There were some concerns about environmental pollution, but these worries did not result in any organized opposition. Local residents were familiar with coal production and use. In other nonmining areas where residents did not have this experience, there was much more concern about the environmental risks of coal-fired generation. The community's familiarity with mining makes the Matsushima case very different from other initial sitings and quite similar to most subsequent sitings, which is crucial in explaining the lack of resistance. General environmental concerns did not translate into consistent levels of opposition to noxious facilities at the local level in Japan.

There was strong mainland support for the project. Ooseto had major economic difficulties, and the local community was attracted to the substantial financial, income, and employment benefits expected from the project. Its geographic location in relation to the project also conditioned the local response. The community thought any environmental costs would be concentrated on the island.

In May 1974, the EPDC explained the details of the proposal to the Matsushima, Ooseto, Taira, and Yukiura fishing cooperatives based in Ooseto town. The little coastal fishing conducted off Matsushima was less important to the cooperative than in Hamaoka. There was some concern about waste water. Most fishermen were engaged in offshore fishing. The oil crisis had increased dramatically their operating costs, and many thought they would gain from transferring their property rights to the utility. Cooperative leaders argued that they should at least allow the EPDC to conduct the EIA so that any adverse effects could be identified and managed. Fishing cooperatives readily gave their consent for the utility to carry out environmental investigations.[25]

There was growing bipartisan political support for the project at local and prefectural levels, which differed from nuclear and other conventional plant sitings in Japan. Politicians from both the LDP and the JSP had important constituencies in Ooseto town and in other parts of the prefecture that relied heavily on the continued protection of the coal industry. The oil crisis further reinforced broader prefectural support for the project in welfare terms. Politically, it was more expedient to promote the project in these terms, even though there were other more economically efficient ways of managing the impacts of adjustment, such as providing lump-sum subsidies.[26]

While the oil crisis intensified support for the project, the introduction of the Three Laws in 1974 increased further the expected gains to the community. The Three Laws guaranteed that some of the benefits to the national community from power plants would be directed to regional areas accepting those plants. Power companies were taxed on the amount of electricity sold, and funds were channeled to host and adjacent communities. The establishment of the Three Laws allowed the local leadership to stress the benefits, particularly for infrastructure development, of accepting the project quickly. It was consistent with attaching a welfare policy emphasis to the project.

There was virtually no local resistance to the EPDC project, even though it was developed in a period when there were broader concerns about pollution problems. This unusual feature was attributable to three main factors. First, the local community was familiar with the risks of the project and was very attracted to the social and economic gains from hosting it. Second, there was little concern among fishermen, given the scale of coastal fishing and the structure of the regional fishing industry. Third, the project had strong bipartisan support at the local, prefectural, and national levels for achieving welfare policy objectives. The lack of any organized economic, political, or ideological resistance distinguishes Matsushima fundamentally from most other initial sitings and makes it look more like a subsequent one. Whereas Kyushu Electric viewed it as a NIMBY project, the local (including island and mainland) and prefectural communities regarded it as a YIMBY (yes in my backyard) project.

External economic and institutional changes can alter community cost-benefit perceptions of noxious facilities, but the extent to which promoters can capitalize on changed expectations is influenced by the perceived legitimacy of bargaining strategies. The oil crisis and the introduction of the Three Laws increased the benefits of the project to both local and regional stakeholders and enhanced further their political support for it. Given the need to obtain agreement from extra-local interests, the EPDC carefully legitimized its approach by promising to wait for results of an impact study before negotiating. This stance maintained and strengthened community support in its political battle with Kyushu Electric and MOF. As we have seen in earlier cases, the perceived legitimacy of bargaining strategies can influence siting outcomes profoundly.

INTERMARKET AND INTERJURISDICTIONAL BARGAINING

As soon as the EPDC obtained consent for the impact study, Kyushu Electric began to oppose the project publicly. Whereas Kyushu Electric expected to be in a favorable supply situation, the western electricity sphere comprising the Kyushu, Chugoku, and Shikoku power companies did not. Supply capacity, particularly in Chugoku Electric's sphere, had not kept pace with demand increases. MITI expected that, if no additional plants began operations by 1979, there would be an electricity shortfall of about 850 mw. It now viewed the project as important in balancing the broader regional electricity market.[27]

Until early 1974, the EPDC had been negotiating bilaterally with Kyushu Electric over the project's construction. The different excess supply situations in Kyushu and in the western electricity sphere provided an opportunity to resolve the conflict by supplying electricity to the western sphere. The EPDC and MITI brought Chugoku and Shikoku Electric into the siting process. They argued the case for supporting the project in terms of achieving welfare, market, and security objectives simultaneously.[28] Chugoku and

Shikoku Electric were supportive because they could not develop large projects economically, as their electricity grids were small. They wanted to be able to benefit from purchasing electricity from other utilities to prevent shortages. Their support exerted additional pressure on Kyushu Electric and enhanced the EPDC's bargaining position.[29]

Existing transmission lines could transmit only small amounts of electricity from Kyushu. Two major issues had to be resolved in the Western Electricity Deliberative Council (Electricity Council).[30] The first involved the number of plants the EPDC would construct and the proportion of electricity each utility would purchase. The EPDC wanted to build two projects, but Kyushu Electric argued that one project would be sufficient to meet regional supply objectives. The Electricity Council decided to support one project, although it did not preclude an additional plant should market conditions justify it. The project had a capacity of 500 mw, and each utility agreed to purchase quantities based on market need and grid size.[31]

The second issue was the burden each company would bear for constructing the distribution network, which was expected to cost about 11,700 million yen. Chugoku Electric agreed to pay 4,600 million yen for transmission lines through Hiroshima and Okayama prefectures. The EPDC agreed to provide 6,100 million yen for all of the other network facilities.[32] Together with Chugoku Electric, it provided indirect compensation to Kyushu Electric by subsidizing the distribution network. The network reduced the potential for an excess supply situation and improved the reliability of the grid both in Kyushu Electric's and in the broader regional sphere. Kyushu Electric incurred no capital costs for the improved distribution network.

Even though the conflict with Kyushu Electric was over, the EPDC still had to get MOF approval for project finances. It appealed to MITI. The Public Utilities Bureau was interested in the project because of heightened concern about energy security. The Coal Bureau in the Agency for Natural Resources and Energy (ANRE), which is a part of MITI, also promoted the project to provide an outlet for domestically produced coal. The coincidence of interests between these two bureaus, despite different objectives, enabled MITI to present a united front in persuading MOF to finance the project.[33]

In late 1974, the EPDC, with MITI backing, formally requested MOF approval for 200 million yen to conduct an impact study and the other preliminary investigations.[34] To the company, getting these funds was a matter of urgency. It wanted to complete the impact study and begin community negotiations so that it could present MOF with a siting proposal in the latter half of 1975, to be approved in the 1976 budget.

The powerful Budget Bureau in MOF was not enthusiastic. It judged the project to be against macroeconomic policy goals.[35] It argued that MITI was already providing adequate subsidies to protect the coal industry and that additional funds were not justified. The ministry was attempting to formulate a budget that reduced the national deficit by restraining public spending. It wanted to reduce aggregate demand in order to curb spiraling inflation that had been exacerbated by the oil crisis. MOF did not approve the financing.[36] Its strategy, like that of anti-siting movements, was to delay the project in the hope of increasing uncertainty and killing it.

MITI stepped into the breach by providing a special subsidy to the EPDC so the company could start negotiations. It stressed the importance of energy security goals and the role that utility investment played in regional economic recovery. MITI mobilized strong local and prefectural bipartisan political support, which exerted considerable pressure on MOF to release the financing. MITI also took the dispute into the national political arena. It persuaded both LDP and JSP politicians, whose electoral bases were in coal-producing areas, to take the matter to the Diet. Many politicians argued that, given nuclear siting difficulties, it was odd to reject projects supported by regional communities. Not even MOF's powerful Budget Bureau could counter strong local, prefectural, and national political pressure, and it reluctantly approved the financing.[37]

In contrast to the other disputes examined in this book, both the power industry and the state had been deeply divided and in conflict with each other over the siting of the project. The oil crisis revised expectations about the worth of the project and provided the catalyst for the conflicting parties to reach compromises. It allowed the EPDC and MITI to provide compensation to Kyushu Electric to reduce its uncertainties about the project's impacts. Citizens were not the only ones who could be compensated for unwanted projects. It also permitted promoters to muster considerable political and bureaucratic support from different levels of government to pressure MOF into financing the project. Changed expectations created a window of opportunity for resolving these extra-local conflicts, but promoters still had to execute skillful strategies, such as bringing opposing interests into regional decision-making bodies, to break the stalemate.

THE DYNAMICS OF COMPENSATION PROCESSES

The success in dealing with Kyushu Electric and MOF created difficulties for the EPDC in its bargaining with the regional community. Negotiations between MOF and the company over project financing were seen by some community interests as breaking the company's promise of waiting for the community's response to the impact statement before deciding on construction. Furthermore, since the middle of 1975, the company had been conducting land investigations, which the community presumed meant the project would proceed. These events injected instabilities into the siting process.[38] Some residents on Matsushima island started opposing the facility to improve their bargaining positions. Not withstanding benefits from the Three Laws, they wanted compensation directly from the EPDC in return for their agreement. They justified their positions by arguing the company had broken its promise over the conduct of the EIA.

In September 1975, a group led by Kamoura Fusahiko, an influential resident of Uchiura, formed the Life Deliberative Council to oppose the project.[39] Kamoura had been a fisherman but, after his father's death, had left the island to study architecture. He later returned to Matsushima and lived a "Robinson Crusoe" style of existence in an abandoned mine owned by Matsushima Kosan. Kamoura and five other people had signed contracts with Matsushima Kosan allowing them to live on the mine site. They were upset with the EPDC, as company officials had been conducting land investigations without their consent. They had a legitimate right to claim compensation. As an EPDC official stated,

> Matsushima Kosan had signed contracts with these residents, allowing them to live on the mine site. Although the contracts ended in 1969, the company allowed them to remain living in the area. They did not argue that they possessed formal property rights like in the case of fishing and land rights. Rather they argued that they owned *iriaiken* (customary rights, or rights to traditional access) and that they would be significantly disadvantaged by not receiving compensation. Customary rights come under the rubric of the Compensation Standards. Kamoura made a wrong career choice in studying architecture. He would have done much better as a lawyer.[40]

This opposition influenced attitudes on the project. Other community interests started demanding a share of the benefits over and above what was available under the Three Laws. They argued that they would incur all of the environmental costs of the project while communities in Honshu and Shikoku would receive all of the gains.[41] Fishing cooperatives in Ooseto town, though not developing any organized opposition, asked the utility to pay adequate compensation. A major conclusion of the impact study was that discharged waste water would affect some coastal fishing. Fishermen knew the utility required their consent and attempted to improve their bargaining position.[42]

Promoters became worried that intense bargaining by some groups to increase their share of compensation might lead to the emergence of organized opposition if the company could not satisfy all competing claims. After all, the company's credibility had come into question. The EPDC responded by conducting a series of discussions in Ooseto to ascertain community attitudes and demands, much like Chubu Electric had in the

Hamaoka dispute. These investigations revealed that community interests were in favor of the project provided their welfare, as they perceived it, would be enhanced. This finding implied adequate compensation.

The EPDC, in conjunction with the town, started negotiations on the distribution of compensation under the Three Laws in early 1976. These negotiations centered on the four villages that amalgamated to form Ooseto town (Ooseto, Matsushima, Yukiura, and Taira villages). The town proposed a plan that envisaged a substantial proportion of compensation going to Ooseto town (28 percent) and to the former Ooseto village (27 percent). The relatively large share for the former Ooseto village reflected its proximity to the project. In contrast, the former Matsushima village and the fishing industry were to receive 14 and 9 percent, respectively. The town felt that these two groups would receive substantial benefits from the transfer of their property rights. The remainder of the compensation was to be directed to the former Yukiura and Taira villages.

The planned distribution of compensation did not reflect the allocation of bargaining power. Matsushima and the fishing cooperatives opposed it. They had property rights and increased their share of compensation at the expense of the town community. The town's share declined from 28 percent to 11 percent. The former Ooseto village's share declined from 27 percent to nothing. In contrast, Matsushima's share increased by 23 percent to 37 percent. The fishing cooperative's share increased by 28 percent to 37 percent. The share going to the former Yukiura and Taira villages declined marginally.[43] Matsushima and the fishing cooperatives negotiated about 74 percent of compensation from the Three Laws.

The company also provided nonmonetary compensation and mitigated against risk. It initially agreed to build a medical clinic and a public hall on the island. Further negotiations resulted in the company consenting to construct a ferry terminal to ease transportation between the island and the mainland. The company also announced a policy of employing half of the construction workforce from the local island population.[44] In short, most of the community compensation was directed to Matsushima island and fishing interests, and not to Ooseto town.

The company also made design changes to mitigate against risk. As the Uchiura port was expected to become congested with coal vessels, it agreed to build a substitute fishing port to reduce the probability of accidents. It also decided to change the positioning of waste water outlet pipes. As fishermen used the area to the west of the island for coastal fishing, the pipes were changed to point east.[45] These cost-reducing, risk mitigation strategies were just as important as those that increased benefits. This dispute and the one at Ashihama demonstrate that community compensation strategies are not likely to facilitate conflict resolution unless they are heavily directed at interest groups with veto power.

The EPDC conducted these negotiations simultaneously with land rights negotiations. Kamoura's opposition destabilized the land-negotiating process. He persuaded other landowners not to sell their land or to demand large payments. To isolate the majority of landowners from Kamoura, the town formed the Landowners Committee in June 1976. Persistent opposition by Kamoura, however, prevented the committee from acting as a single negotiating body. Prefectural, town, and EPDC officials visited Kamoura to persuade him to accept the project. He continually opposed it. The EPDC decided to increase the offer price for land transfer and to negotiate individually with landowners. During July and August, it completed the majority of land negotiations.[46]

In October, the EPDC started negotiations with the fishing cooperatives. It offered 420 million yen to the Matsushima, 98 million yen to the Taira, and 75 million yen to the Yukiura fishing cooperatives. Although the fishing cooperatives had given their in-principle approval they rejected these offers as being too low. They perceived that their bargaining positions were now much stronger because the EPDC had nearly finalized community and land rights transfer negotiations. They knew that failure to win their approval would substantially delay or even cause abandonment of the project.[47] Fishing cooperatives were not only tough but also clever bargainers and requested more compensation than in the case of most nuclear power plants. The EPDC relented and received EPDCC approval in late December 1976.

The sudden upsurge of resistance to the project had the potential to become organized resistance. The EPDC had two choices in responding to the opposition. The first was to respond like Chubu Electric in the Ashihama case. The EPDC, as a government-owned company, could have invoked the state's eminent domain powers to force property rights transfer. With concern about oil supplies, this reaction could easily have been portrayed to be in the national interest. The second choice was to respond like Chubu Electric did in the Hamaoka case—negotiating with community interests. The EPDC chose the latter. In this sense, the state also learned from the Ashihama disaster—utilities need to negotiate with resistance rather than trying to bypass it.

Despite virtually no organized resistance, fishing cooperatives extracted large compensation from the EPDC and got the lion's share of benefits under the Three Laws. This victory reflected their relatively strong bargaining positions. Although they were interested in alleviating financial problems, the EPDC had fewer alternatives to negotiating a settlement. Failure at Matsushima would have meant another similar negotiating process and associated uncertainties. Changed expectations increased the compensation pool to win agreement. Fishing cooperatives knew they could demand large compensation outlays because both Kyushu Electric and MOF could start opposing again should negotiating processes become stalled. After all, the

EPDC had lost its political legitimacy at the local level. The Matsushima project cost the state dearly.

CONCLUSIONS

In contrast to Ashihama and Hamaoka, Matsushima was a case of a utility paying compensation to another utility and unorganized interests to site a project. The EPDC paid large sums of indirect compensation to Kyushu Electric (6,100 million yen) to cover the costs of further developing its electricity distribution system. It convinced other utilities to also indirectly compensate Kyushu Electric. Large financial packages were offered to community interests. The EPDC outlaid a much larger amount of compensation to fishing cooperatives (2.3 million yen) than at Ashihama or Hamaoka. It provided risk mitigation and other nonmonetary compensation to other community interests. Unlike the previous two disputes, community interests (mainly fishing cooperatives and Matsushima island residents) also received large funds under the Three Laws (4,250 million yen).

The broad bargaining environment in which the dispute was played out was structured in a way that, according to the earlier quantitative analysis, would have led to a much longer settlement time (sixty-eight months) compared with the average of all initial fossil-fueled power plants in Japan (thirty-eight months). Although there was no need for the project in the Kyushu Electric's sphere, there was some need for it in the western electricity sphere. Incomes were increasing relatively quickly, and the prefectural community wanted to maintain that growth. The financial base of the local government was relatively weak, and the community expected that the project would alleviate financial problems. Leftist political parties, which had relatively strong representation in regional assemblies, emphasized the development of a fossil-fueled project, and social attitudes encouraged preservation of the environment more than economic growth. The favorable electricity situation dominated these other influences and explains why the project took longer than average.

The polimetric analysis provides an important perspective on the structure of the overall environment in which participants managed the conflict. Kyushu Electric's market prospects were favorable, and it opposed the project. Although broader social attitudes placed relatively more weight on environmental preservation, local and prefectural governments and political parties supported the project at all stages of the conflict. Prospects in the rural sector were relatively unfavorable and, unlike the Ashihama and Hamaoka cases, no organized resistance emerged. In short, these observations suggest a bargaining environment where relatively little economic surplus was available for redistribution to opposing interests. Despite strong local and prefectural support, it was not easy for the EPDC to compensate Kyushu Electric, particularly in the early stages of the dispute.

Although the size of the residual in the Matsushima case (twenty-four months) was larger than in the Hamaoka case, the model overestimated the time required to resolve the conflict. Understanding this short resolution time requires exploring how external shocks influenced the structure of the overall bargaining environment. Changed market and institutional circumstances increased the surplus that the utility could redistribute to community interests with veto power. The added surplus was crucial in striking a deal with Kyushu Electric but this meant the EPDC breaking its promise to the local community. The credibility of the company's approach then came into question. Community interests exercized their bargaining power to reap very large payments despite not being organized. The potential threat of organized resistance was strong enough the force the EPDC to compensate handsomely.

The major costs perceived from the project fell unevenly across community actors. Unlike the Hamaoka case, there was virtually no organized resistance, despite growing environmental concern. The project's geographic location was important in explaining the pattern and intensity of resistance—it was sited on a sparsely populated island, the community had familiarity with coal technology and pollution risks, and there was little coastal fishing in the vicinity of the project. In contrast, strong opposition emerged from another utility and MOF. Kyushu Electric opposed the project because of the expected costs of managing its electricity sphere. MOF opposed because of the expected costs of inefficient welfare policies (and subsidies) given macroeconomic policy priorities. Matsushima was fundamentally different from most disputes involving private utilities. Local and prefectural communities were less divided over the project than the power industry and the state.

The allocation of bargaining power favored local and regional community interests more than it did the EPDC and MITI. Property rights owners had alternatives to striking a bargain with the utility. The political fallout from the utility breaking its promise about conducting the EIA also politically enhanced these alternatives. The utility perceived that it had fewer viable alternatives to negotiating an agreement. Although the oil crisis enhanced the EPDC's bargaining position with Kyushu Electric, it did not improve the EPDC's position with community interests because of the time it would have required to find an alternative site. Even though the EPDC had strong backing from MITI, its status as an "electric nomad" weakened substantially its bargaining position in siting.

Conflicting parties executed a range of bargaining strategies in line with their negotiating positions. The EPDC cleverly brought Kyushu Electric into a broader regional decision-making body to exert pressure on it. Both the EPDC and MITI successfully marshaled bipartisan political support at all levels of government to pressure MOF into financing the project. Its strategies in dealing with private and bureaucratic opposition contradicted its approach to negotiating with local interests. The loss of the EPDC's credibility enhanced the negotiating strengths of local interests. They effectively extracted large compensation payouts for their consent.

Expectations about the value of the project changed fundamentally during the course of the dispute. The oil crisis and the changed economic and institutional environment increased the expected value of the project to both proponents and opponents. It allowed the EPDC and MITI, through carefully crafted strategies in a changed bargaining environment, to negotiate a deal with Kyushu Electric and MOF. These negotiations created instabilities at the local level. Community interests got more compensation from the EPDC, even though the Three Laws provided considerably expanded compensation packages. A local official provided a useful insight into the changed positions of community interests: "If the EPDC is going to provide lots of money to Kyushu Electric, which is opposing publicly, then perhaps we should up our opposition so that we make sure we get a fair deal."[48]

Like other siting cases, high levels of uncertainties beset conflicting parties initially. In the early stages of the dispute, the major uncertainty related to managing Kyushu Electric's opposition. Interestingly, concern about the environmental risks was not a major feature of the dispute. Although uncertainty increased during the conflict, the oil crisis ultimately enabled the EPDC to manage those uncertainties, by compensating (together with other private utilities) Kyushu Electric for adverse impacts of the project on its electricity sphere and negotiating a deal with MOF over project financing. These processes raised community uncertainty. The company offered very large compensation outlays to reduce this uncertainty to the point where a deal could be reached.

Changed expectations, on the part of a least one participant, are necessary but not sufficient conditions for resolving siting conflicts. Maintaining credibility in responding to changes in the negotiating environment is crucial if those changes are to facilitate conflict resolution. The Matsushima case, like the Hamaoka case, provides further evidence that NIMBY conflicts can be resolved reasonably quickly. Although the nature and positions of the conflicting parties were different from other cases examined in this book, bargaining and the use of compensation did bring conflicting parties together to reach a siting agreement.

CHAPTER SEVEN

Dealing with Changing Project Costs

Siting risky projects involves deciding who wins and who loses from developing those projects. Where communities have veto power, negotiating agreements requires developers to compensate community interests for losses expected to be incurred from facilities. A crucial element is whether developers expect there to be net private gains after bargaining and construction costs have been taken into account. Where they expect net losses, they may reschedule the project and develop other, less costly ones. Relative bargaining and construction costs of competing projects will influence approaches to scheduling negotiations at alternative sites.

The Tomari case is important because it highlights siting difficulties encountered in siting initial nuclear projects, which came about during the 1980s. It took the utility about thirteen years to win community approval for the project, considerably longer than the average time of about eight years for all initial nuclear projects in Japan. During the 1980s and 1990s, almost all of Japan's nuclear capacity expansion came from developing subsequent projects, which had significantly shorter bargaining times. The few approved initial nuclear projects, including Tomari, all had approval times of greater than twelve years. Comparing the Tomari dispute with the more easily resolved (Hamaoka and Matsushima) and the most difficult ones (Ashihama) completes our empirical analysis of bargaining and compensation.

Tomari is different comparatively in that very large amounts of compensation (similar to Matsushima) were required to resolve a dispute that resulted in one of the longest approval times in the history of nuclear siting in

Japan. Until the oil crisis, Hokkaido Electric was not prepared to compensate regional fishing cooperatives, as there was no real economic justification for developing the project. The oil shock broke the impasse, but the prospect of a terrorist attack made the project unviable. The utility changed the project's location, which weakened the power of opposing fishing interests and enabled the company to negotiate a settlement. Bargaining is delayed not only by opposition; it can also be delayed intentionally by developers if they face high or rising project costs.

SITE SELECTION AND RESISTANCE

In late September 1969, Hokkaido Electric, together with the Hokkaido government and the Sapporo branch of the Ministry of International Trade and Industry (MITI), decided officially to develop the Kyowa-Tomari nuclear plant. The company expected to start construction in 1973 and operations in 1977.[1] The relatively long planning horizon (about four years to gain community acceptance) was directly attributable to the utility's motivation for developing the project. It saw the project as focused on technological research, not as an economically viable option for balancing the electricity market in Hokkaido.[2] Since 1967, MITI had been considering candidate sites and had proposed subsidies to Hokkaido Electric. The utility was eager to develop a nuclear project so as not to be left behind technologically. Like Chubu Electric at Ashihama, a major reason for starting the siting process was to obtain national government subsidies for technology development.

There was no strong economic justification for developing the project. In 1970, the utility expected to have an oversupply of approximately 560 megawatts (mw) by 1975. The speedy development of the Kyowa-Tomari plant would have resulted in too much excess capacity (40 percent).[3] A relatively small grid prevented development of a large plant to reap economies of scale.[4] Furthermore, the Hokkaido grid was interconnected with the Honshu grid by only a small transmission line, and the company did not see any prospects for selling excess electricity to Honshu. Company officials believed the project would reduce the reliability of the grid in the event of an outage.[5]

To improve the project's viability, the power company sought to minimize construction costs.[6] Its low budget resulted in a very unusual plant design. In 1968, MITI and the Hokkaido government selected Tomari and Hamaeki villages as candidate sites. Hokkaido Electric also conducted site investigations at Kyowa village. Although there were some political considerations, economic factors dominated selection of the Kyowa-Tomari site. Hamaeki village was too far away from electricity consumption areas in central Hokkaido and would need transmission lines and improved port facilities. The Kyowa and Tomari sites were closer to consumption areas, but location in Tomari would have required substantial excavation. As shown in map 5, the Kyowa site, although two kilometers inland, offered the least costly siting package. The town had already purchased land for the project. The land required no excavation and was close to good transportation routes to and from the large port at Iwanai, located to the west of Kyowa town.

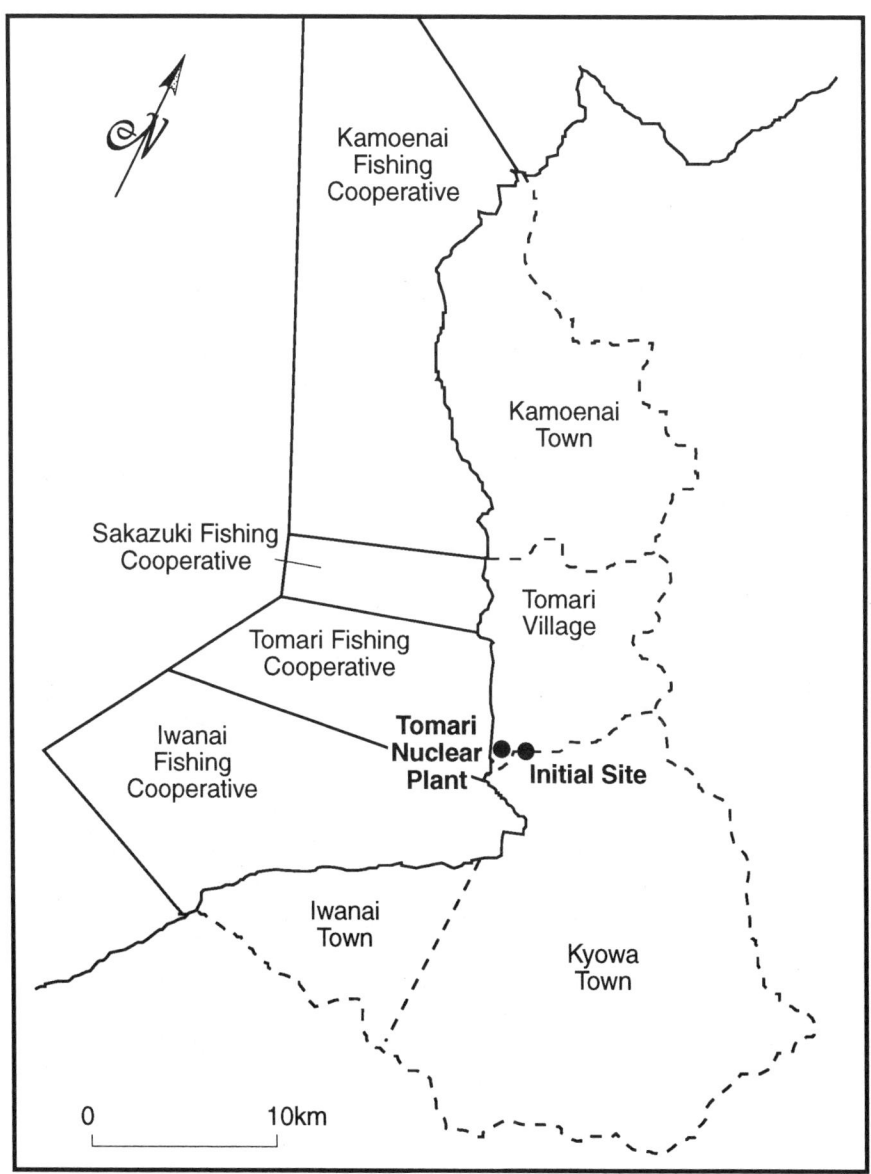

Map 5 Tomari nuclear plant: site location and property rights ownership

Although the containment vessel was to be located in Kyowa town, the company did not wish to lay waste water pipes to the southwest, the most direct ocean route, because of anticipated resistance by farmers. Farmers in Kyowa used the Karifuto river as a source of water. As they had water usage rights, the company expected it would have to pay compensation to them. The utility decided to lay the waste water pipes underground through Tomari village to the ocean. This design, with exceedingly long waste water pipes, although never before attempted in Japan, would allow the village to receive a share of fixed-asset taxes. The company did not expect any problems in dealing with farmers in Tomari. Any resistance from Tomari, which might arise if Kyowa were to receive all of the benefits, would be mitigated. This plan would enable the benefits to be more evenly distributed than if the plant had been located solely in Kyowa.

On 30 September 1969, local mayors of Kyowa and Tomari villages and Iwanai town signed memorandums with Hokkaido Electric promising cooperation in siting the plant. The first signs of organized resistance came from Iwanai fishermen who established the Policy Committee for Opposing the Construction of a Nuclear Power Plant (Policy Committee) in late March 1970.[7] The Iwanai fishing cooperative had fishing rights stretching from Iwanai port to the border of Kyowa town. Even though the Iwanai cooperative did not have fishing rights in the vicinity of the pipes, the utility needed to negotiate for the use of its rights to use the port for transporting components and nuclear fuels. As a local fisherman put it, "The utility and local Iwanai leaders neither consulted us over their plans for the nuclear project nor their plans to use the Iwanai port. This was totally unacceptable. Apparently, Hokkaido Electric and the leadership got a real shock when they learned that legally we had property rights and that they would have to deal with us. They did not realize that we would have such an impact on siting the project for simply using our port."[8]

In Iwanai town, there were prospects for continued expansion of the fishing industry. Between 1965 and 1971, the value of fishing had improved from 160 to 300 million yen, and the cooperative anticipated further increases to 600 million yen in the next five years. The most important types of fish to the cooperative were walleye pollack caught near the coast and salmon trout caught in Soviet territorial waters. Fish processed by the cooperative was worth about 400 million yen annually. It was the major processing and distribution center for the industry in the Ganu region, consisting of the Tomari and Sakazuki cooperatives in Tomari village and the large Kamoenai cooperative in Kamoenai village.[9]

The Policy Committee formed an economic alliance with other cooperatives in the Ganu region to improve its bargaining position. On 25 June 1970, it established the Fishing Alliance for Opposing the Construction of the Nuclear Plant (Fishing Alliance). It sponsored a series of lectures on nuclear safety and ocean pollution in Iwanai, aimed at creating more local awareness of nuclear risks.

Despite early signs of strong collective action and the use of protest as a political resource, the Fishing Alliance was short-lived. The expected negative impacts of the project differed across the four cooperatives. The Kamoenai cooperative did not have property rights in the area where waste water was to be discharged. The smaller Sakazuki and Tomari cooperatives were heavily involved in offshore fishing and did not expect to be affected adversely. The Sakazuki cooperative, which actively promoted the project, had the smallest coastal fishing sector in the region and almost all of its catch was derived from offshore fishing. The Tomari cooperative decided to undertake a waste water investigation, as a higher proportion of its catch originated from coastal fishing.[10]

The ability of the Fishing Alliance in Ganu to oppose development was weaker than that of Nanto fishermen at Ashihama, although stronger than that of Hainan fishermen at Hamaoka. The size of the industry in Ganu was smaller than in Nanto but larger than in Hainan. The adverse effects of the Ashihama project were evenly distributed among pearl producers in Nanto. In contrast, as in Hainan, the effects of the project did not fall evenly across cooperatives in Ganu. For example, the Sakazuki cooperative promoted the project because deep-sea fishermen expected to get compensation for coastal fishing rights. The disproportionate spread of the expected burden of the project weakened the alliance's collective opposition.

An anti-nuclear movement, which called itself the Five Member Struggle Committee (Struggle Committee), emerged. It consisted mainly of the Japan Socialist Party (JSP) and its affiliates, which were based outside of the Ganu region. The movement was concerned about nuclear risks, which reflected rising concerns about the quality of the environment in the late 1960s and early 1970s.[11] It was particularly worried about the failure of an Emergency Central Cooling System (ECCS) experiment in the United States and the Tsuruga nuclear accident in Japan, both of which led to considerable distrust of Japan's nuclear administration in the early 1970s.

The movement had another reason to oppose nuclear energy. Hokkaido, like Kyushu, was a major coal-producing area. The Hokkaido branch of the JSP promoted the protection of the coal industry. Steaming coal production in Hokkaido had declined from fifteen to ten million tonnes between 1965 and 1971. There was concern that production would decline further after the national government decided in 1968 to phase out domestic coal. To the Hokkaido JSP, nuclear power represented a further move away from coal use, which would have unacceptable social and economic consequences.[12]

The Struggle Committee attempted to sensitize the local community to nuclear risks. Beginning in early 1971, it sponsored a series of meetings in Iwanai and sent petitions to the power company, the prefectural assembly, and the four towns and villages in the Ganu region. The movement mustered some support for its position in Iwanai, but it was unable to do so at Kyowa town and Tomari village, predominantly rural areas dominated by conservative politicians. These two towns supported the project for regional development purposes. As in Hamaoka, there was virtually no economic, political, or ideological base on which to stimulate community resistance at the local or the regional level.

In contrast, at the prefectural level, the movement was quite successful in appealing to the Liberal Democratic Party (LDP). Like the JSP, the LDP had a policy of protecting the domestic coal industry. The party leadership was worried that strong support for the project might create intra-party conflict. The JSP's strength in the Hokkaido assembly had increased to a little less than half of the seats, and there was a risk that the JSP would capitalize on such conflict in the coming prefectural election.[13] Hence, the prefectural government was not willing to play a major role in supporting the project openly and in mediating between the utility and the Iwanai cooperative. This ambivalence contrasts with the Matsushima dispute, where bipartisan support for the coal-fired project facilitated conflict resolution. These two examples provide some evidence that prefectural governments in coal-producing areas are more likely to support coal-fired than nuclear projects.

Hokkaido Electric's interest in developing the nuclear project was to obtain subsidies to develop nuclear technology and not to balance electricity markets. The economic, ideological, and political resistance that emerged exacerbated uncertainties for the power company. The uneven spread of costs, especially among fishing interests, affected these interests' ability to sustain opposition. There was the lack of support from the prefectural government because of bipartisan support for protecting the coal industry. It was, however, the company's unwillingness to negotiate the siting of an uneconomic project, rather than the strength of community resistance, that was more important in delaying the project during the early stages of the dispute.

TOWN RESPONSE AND RESCHEDULING PROJECT SITING

Despite the emergence of unexpected opposition, the mayor of Iwanai, Nagahama Toshio, was favorably disposed to the plant. Although Iwanai was the biggest town in the Ganu region, its economy was facing several problems. The town had become more reliant on national government financing. Nagahama was worried about the lack of social overhead capital. He believed that the project would stimulate growth in Iwanai and the regional economy. He quickly received business and political support for the project from the Iwanai Chamber of Commerce and Industry as well as the Group for the Promotion of Nuclear Power, consisting of local conservative politicians.[14] As in Hamaoka, promoters sought to reduce safety concerns by using the demonstration effect. They sponsored lectures on nuclear safety and took community groups to visit areas in Honshu where nuclear plants were being developed or were in operation.

The views of both opponents and supporters of the project were registered with the Standing Committee for Iwanai Town Industry (Standing Committee). The town assembly formed this committee in 1969 to consider and evaluate alternative development strategies for the town. It consisted of local politicians who represented the major interests in the dispute. By early 1972, it was clear that the committee was in gridlock. The town had become split between proponents in the service and construction sectors and opponents in the fishing sector and labor movement. Opposition forces were gaining strength, and so the Standing Committee decided to oppose the plant on 24 June 1972.[15]

Nagahama did not succeed in reversing the decision. On 20 February 1973, he froze the memorandum the town had signed with Hokkaido Electric in 1969. Nagahama was an independent candidate and had won the 1969 election, uncontested, with the backing of the Regional Council of Trade Unions and the Iwanai fishing cooperative. Their support was critical to his position as mayor. It was clear to him that continued promotion of the project, although in the interests of the commercial sector, would result in loss of support from fishermen and unions in the coming 1973 elections.[16] These developments also worried Hokkaido Electric, as it did not want the nuclear issue getting caught up in electoral politics, as it had in Ashihama.

Subsequent discussions between the company and the mayor led to a declaration that contained three major points. First, the town would oppose the plant until there was agreement over nuclear safety. Second, the town would examine a new memorandum with Hokkaido Electric if the power company could guarantee nuclear safety. Third, the town would continue with its plan for expanding the Iwanai port but would not construct facilities related to the construction and operation of the nuclear plant as long as fishermen opposed the project.[17] As a close aide recalled, "Nagahama's position represented a skillful compromise between the opponents and supporters of the project in order to maintain his electoral backing as an independent politician. He could not afford to lose support from either the proponents or opponents of the project. He needed to be seen to be delaying or opposing the project at that time, but also to be leaving open the option for its development in the future."[18] This delaying tactic did not mean that Nagahama had ended his support for the project. Rather, it was a strategy for regrouping if conditions became more favorable.

Freezing the memorandum put the utility in a difficult position with respect to its electricity market. By 1973, it was expecting a shortfall on five-year forecast demand. Between 1970 and 1973, forecast electricity demand had increased from 2,240 to 2,820 mw. The Tokoma-Atsuma fossil-fueled plant, which was being licensed, was expected to add 350 mw to the grid. The company expected a shortfall of about 500 mw.[19] The only two large plants that could conceivably fill this gap were the Kyowa-Tomari nuclear plant and the Date oil-fired plant.

Hokkaido Electric had been negotiating with Date city over the development of the oil-fired plant.[20] Fishing cooperatives in Date city strongly opposed it. The major problem was racial conflict between the Date fishing cooperative, consisting of Japanese, and the Ainu-dominated Usu fishing cooperative. Legally, Hokkaido Electric had to purchase only fishing rights from the Date cooperative. It was prepared to pay only cooperation money to the neighboring Usu cooperative. The Ainu, with the support of the JSP, argued that Hokkaido Electric and the Date administration were not taking the Ainu's interests into account as they were also incurring risks from the project and claimed more compensation. The conflict became a prefectural political issue in the early 1970s.

The power company, faced with substantial resistance at both sites, decided to give priority to the Date plant. It postponed the nuclear plant by two years and planned to start construction in 1976. This scheduling strategy reflected the increased economic and political costs. Initial construction cost estimates had increased from 25,000 to 35,000 million yen in early 1973. In contrast, the construction cost of the Date project, which comprised two plants, totaled 40,000 million yen. Hokkaido Electric was particularly worried about the potential impact of higher construction costs on already increasing electricity tariffs in Hokkaido.

The bargaining costs were also higher for the nuclear, as opposed to the oil-fired plant. The power company's negotiating position was weaker with the community in Ganu than in Date. The utility had received information that the Ganu cooperatives would demand cumulatively about 2,500 million yen for the transfer of property rights. This demand was based on a claim by the Onagawa cooperative over Tohoku Electric's Onagawa nuclear project. Hokkaido Electric expected such a payment to increase the cost of the nuclear project by about 9 percent. In contrast the Date and Usu cooperatives were requesting a total of about 900 million yen in payment. The power company calculated that this would increase the cost of the oil-fired project by 2.5 percent. The utility also expected negotiations to take longer for the Kyowa-Tomari nuclear plant, given the Iwanai assembly's anti-nuclear resolution. Hokkaido Electric had already started negotiations with Date fishing groups. Furthermore, Date city was promoting the plant and was acting as a mediator in the dispute. The doors to bargaining were open in Date but not in Iwanai.

Changed bargaining conditions influenced the dispute in different ways. Promoters managed these changes not by trying to pressure local resistance like in Ashihama but, like at Hamaoka, by using alternative strategies that were perceived as being more legitimate and that kept open the possibility of bargaining. Changed market conditions increased the company's compensation pool, which meant very little as both the town and the prefecture were not actively supporting the project. The utility had the cash but not the mediators. In averting electricity shortages, it distanced itself from the Iwanai dispute and pushed ahead with the Date oil-fired project. There was less uncertainty at the latter site, as the utility had mediators to assist in negotiations with fishing cooperatives.

REVISED EXPECTATIONS AND BARGAINING STRATEGIES

Although the utility originally planned to postpone the Kyowa-Tomari plant by about two years, it was not until 1976, four years later, that the company seriously moved forward again with it. During the period from 1974 to 1977, actual electricity demand in Hokkaido increased by 11 percent per annum. This increase was attributable to an upswing in economic activity in Hokkaido and the development of large industries, such as the motor car industry in southern Hokkaido. Electricity supply increased at only 3 percent annually, however, primarily because of delays in fishing negotiations over of the Date plant. The changed electricity market situation again improved the value of the project to Hokkaido Electric.

Its importance also grew because of heightened concern about supply security. Until the early 1970s, Hokkaido's electricity supply capacity consisted mainly of coal-fired and hydroelectric power. The Date oil-fired plant was important in Hokkaido Electric's policy of diversifying away from domestic coal. After the oil crisis, energy policy in Hokkaido stressed further diversification into nuclear power.[21] The utility argued that the need for a nuclear plant was greater in Hokkaido compared to other parts of Japan because of its greater vulnerability to electricity supply disruptions. The Honshu electricity grid was well-interconnected. In contrast, Hokkaido was connected to the mainland by only a small transmission line. The utility could not rely on purchasing electricity from other utilities should electricity shortages occur. The rising uncertainly about oil price and supply enhanced perception of a need for nuclear power.

Policy changes at the national level also had significant impact on regional bargaining processes. The establishment of the Three Laws increased the expected benefits of the project to the community. Kyowa town expected to receive about 1,000 million yen whereas Iwanai, Tomari, and Kamoenai each expected to obtain about one-third of that amount.[22] Their local governments created community expectations that the project would bring social and economic benefits quickly. Their leadership was concerned that resistance might emerge if these benefits were not realized.[23]

Changes in the nuclear administration also reduced the perceived risk of nuclear power to the regional community. Previously, the promotion and regulation roles of the nuclear administration were located in one body, which had reduced the credibility of the administration. The Science and Technology Agency split these roles and established the Nuclear Energy Committee and the Nuclear Energy Safety Committee. The government also instituted a system to double-check nuclear safety. In these ways, the national government reduced the concern in the community about risks.

These economic and policy changes facilitated bargaining by enlarging the compensation pool (increasing perceived economic benefits and reducing perceived risks) that could be used in negotiation. The renewed and stronger regional interest in the project put pressure on Iwanai town and the Iwanai fishing cooperative to accept the nuclear plant. Other areas to the north of Iwanai argued that the benefits of the project had increased while the risks had been reduced. Furthermore, they believed that continued opposition in Iwanai would be at the expense of important regional development objectives.

Yet the major stumbling block was opposition by the Iwanai cooperative. The scope of the cooperative's fishing rights extended to the use of the port. As these rights conferred "exclusive use" of the port, the utility had to negotiate with the cooperative. This veto power put the cooperative in strong bargaining position. The company attempted to isolate the Iwanai resistance. Its strategy was to commence negotiations with the Tomari cooperative and then attempt to sign a new memorandum with Iwanai town. It believed that success on these two fronts would improve its bargaining position with Iwanai fishing interests.[24]

In 1974, the utility began to negotiate with the Tomari cooperative. Initially, as in other disputes, there was a substantial gap between what it was prepared to pay and what the fishing cooperative was prepared to accept. The utility based its offer on the Compensation Standards. The Tomari fishing cooperative based its price on the amount being demanded by the Onagawa fishing cooperative. Hokkaido Electric's negotiators thought the utility's position with the Tomari cooperative was relatively weak. The Iwanai cooperative's bargaining position would be greatly enhanced if the Tomari cooperative opposed the project.[25]

Like the EPDC at Matsushima, Hokkaido Electric was constrained in site selection. As it could locate plants only in Hokkaido, it could not induce competition among different prefectures to provide an alternative site.[26] It considered other locations within Hokkaido but assessed that they would be more costly. Cooling water was not available all year round in northern Hokkaido, and the cost of transmission lines to southern Hokkaido prevented siting in eastern Hokkaido. This lack of alternative sites also weakened the utility's bargaining position.

Hokkaido Electric, in negotiating with the Ganu cooperatives, also had to consider other siting negotiations. A major concern of the power industry in the 1970s was increased compensation claims for developing nuclear plants. As fishing cooperatives relied more on the Compensation Standards' "similar situation formula" in transfer negotiations, the industry perceived that compensation claims would increase as each fishing cooperative would base its demands on the highest claim made previously. There was grave concern that resolving the Onagawa dispute could reduce the economic viability of nuclear power in Japan.[27] The power industry was exerting considerable pressure on Tohoku Electric not to pay large sums to the Onagawa fishing cooperative. Hokkaido Electric was also criticized in its negotiations with the Date fishing cooperatives. This criticism prevented the utility from paying high amounts of compensation to both the Tomari and the Iwanai fishing cooperatives. Private utilities not only dislike public companies encroaching on their turf but also dislike other private companies negotiating in ways that influence adversely bargaining on their own turf.

Hokkaido Electric faced a dilemma in negotiating with the Tomari cooperative. It was in a weak bargaining position, as alternative nuclear sites were very scarce, and it was constrained from offering overly large compensation. It negotiated first on fishing development compensation and left property rights transfer arrangements for later. On 9 October 1976, Hokkaido Electric signed an agreement with the Tomari cooperative agreeing to pay ninety million yen for fishing development after the conclusion of property rights transfer arrangements.[28] This deal attained three objectives simultaneously. First, it kept the Tomari cooperative at the negotiating table. Second, it improved its position with the Iwanai cooperative. Third, it minimized the negative impact on siting negotiations at other sites.

The utility had earlier persuaded the prefecture, in conjunction with the Tomari cooperative, to undertake a marine impact study. Its major conclusion was that the negative impacts of the project would be greater on the Iwanai fishing cooperative than on other cooperatives in the region. This finding stemmed from temperature increases in areas used by walleye pollack as a breeding ground.[29] In response, the utility changed the location of the waste water pipes two kilometers to the north and extended the mouth of the pipes a kilometer out to sea, thereby minimizing the impact of waste water on the fishing industry. This strategy was a technical fix to a political and economic bargaining problem.[30]

The utility's position was also enhanced by the increase in oil prices after the oil crisis. Higher operating costs caused financial problems for deep-sea fishermen.[31] In 1976, they created the Committee for Considering Nuclear Energy and the Fishing Industry (Fishing Industry Committee). The committee commenced negotiations with Hokkaido Electric over the use of the Iwanai port and compensation for fishing development.[32] Like at Hamaoka, the effects of the power plant on deep-sea fishermen differed from the effects on coastal fishermen. Deep-sea fishermen saw compensation as a means of reducing large debts that they had incurred.

The utility also increased the benefits of the project to Iwanai town over and above those expected to accrue from the Three Laws. In 1977, it bought town land at a cost of 4,000 million yen, which was considerably greater than market rates. These funds helped the town to overcome financial difficulties until compensation from the Three Laws was available. The company also promised to give priority to purchasing goods and employing unskilled labor from the town.[33] The use of considerable indirect compensation further enhanced the benefits of the project to Iwanai town.

The weakening of opposition from fishing cooperatives and the strengthening of community support in Iwanai town allowed Nagahama, on 12 March 1977, to declare his intention to sign a new memorandum with the utility.[34] He campaigned for the July 1977 election promoting the project. Nagahama won the election overwhelmingly over a JSP candidate who opposed the nuclear project. His earlier strategy of withdrawing active support for the project but leaving open the option of developing it at a later time clearly paid off. Nagahama's victory reflected a changed attitude in Iwanai. He skillfully used the company's willingness to address safety concerns and compensate the community to create support for the project.

Although the Fishing Industry Committee had started negotiations with Hokkaido Electric, there was still considerable conflict with coastal fishermen over the distribution and use of compensation. Coastal fishermen argued that the waste water would directly affect walleye pollack breeding grounds despite the change in the location of the waste water pipes. They believed that the deep-sea fishermen would receive a larger share of compensation despite incurring no costs from the project.[35] Deep-sea fishermen were stronger in numbers and, as the distribution of funds from the sale of property rights is decided internally, coastal fishermen were in a weaker position to gain what they considered to be a fair share of compensation. As a local fisherman said, "We (coastal fishermen) were incurring virtually all the costs, yet deep-sea fishermen would receive the bulk of fishing compensation as they had the numbers in the cooperative. We kept on opposing this situation but the leadership of the cooperative just would not listen to our arguments. It simply was not fair."[36] As in many other siting disputes in Japan, there is an asymmetry between who loses the most and who gains the most from fishing rights transfer negotiations.

National policy and administrative changes after the oil shock revised promoter and community expectations about the value of the project. These changed expectations increased the expected benefits and reduced the expected risks of the project. They provided an opportunity for promoters to end the deadlock. Yet that also required carefully designed bargaining strategies. The company's bargaining position was relatively weak. It had very limited siting options and was constrained in paying compensation by other utilities. It managed this by shoring up community support, particularly in Iwanai, through the use of both direct and indirect compensation and by commencing negotiations with more supportive fishing cooperatives. This strategy was crucial in the reelection of the mayor of Iwanai, which provided him with an electoral mandate to play a more active role in supporting the project and in isolating local fishing opposition.

PLANT SITE MODIFICATION

Despite promising developments, other external events injected negative instabilities into bargaining processes. In April 1977, the Soviet Union established a 200-mile economic zone. This zoning had a major impact on the Iwanai cooperative's attitude toward the plant. Forty percent of its catch was now in jeopardy. Deep-sea fishermen who had large debts felt that they would have to move into coastal fishing. The expected value of coastal fishing increased markedly. In March 1978, the Fishing Industry Committee decided to oppose the project.

At the same time, the company was stunned to learn that it was also confronted by the risk of a terrorist assault on the plant. Leftist radicals who bombed the control tower at the Narita International Airport stated that the Kyowa-Tomari nuclear plant would be next on their agenda. This threat caused immediate concern among promoters, who knew that any danger of attack would be costly in terms of both plant operation and management. They were anxious about local opposition if they could not guarantee safety. MITI was particularly concerned that the terrorist issue would become entwined in the safety dispute and impede the development of nuclear power.

Promoters conceded that the project was extremely vulnerable to terrorists. The complex would cover a large area and would require costly surveillance to detect and repel any assault. Two design features were noteworthy. The first was the distance involved in loading materials and components at Iwanai port and transporting them to and from the plant. The second, and perhaps more important, was the extreme vulnerability of the waste water pipes, which stretched approximately 6,000 meters to the ocean.[37] As a company official stated, "Just as we were improving our position to bring the dispute to an end, this should happen. We had no choice but to take their threats seriously. MITI told us that it would not support the project with the current design. We had to change the whole design. As far as I am aware, this was the first time there was a relationship between terrorism and nuclear siting in Japan. Personally speaking, it was the biggest problem we faced in siting the project."[38]

Renewed opposition by Iwanai fishermen and the terrorist threat increased uncertainty about developing and operating the plant, but changed market conditions decreased the need for the project. Although demand was expected to increase to 3,660 mw in the next five years, the utility had sufficient capacity in the pipeline to equilibrate electricity markets. The two Date plants were ready to commence operation. The Tokoma-Atsuma coal-fired and Shiriuchi oil-fired plants had just received Electric Power Development Coordination Committee (EPDCC) approval and were expected to start supplying power within three years. In 1978, Hokkaido Electric decided that it would postpone construction until 1983.[39]

The utility decided to change the project's location to a site solely within Tomari village and build two plants instead of one.[40] Cost calculations suggested that this would save about 4,100 million yen. The major part of this expected savings (3,500 million yen) stemmed from economies of scale derived from constructing two projects. There were other anticipated savings, including those from avoiding port access negotiations at Iwanai (750 million yen) and construction of a special road from Iwanai to Tomari (50 million yen). There were, however, added costs of constructing a port and acquiring land in Tomari (200 million yen). The utility assessed that the net benefits of the site change were even greater than this monetary value if the reduced risk of a terrorist attack were considered.[41]

Although the site change weakened the bargaining position of the Iwanai cooperative, it did not greatly enhance the position of Tomari village. It was already eager to accept the project, expecting to obtain larger benefits than would have been the case under the initial design. These included more fixed-asset taxes and a larger share of the funds from the Three Laws. The utility's position was now relatively strong. Between 1978 and 1980, electricity demand growth tapered off as a result of higher energy prices in Hokkaido arising from the second oil shock. It did not have to develop the project quickly to balance the electricity market in Hokkaido.[42]

Interestingly, there was growing resistance to the site change in Kyowa town. For some time, the community had been looking forward to the economic benefits from the project development. The site change meant that its share of fixed-asset taxes, subsidies from the Three Laws, and other indirect gains would be reduced substantially. Residents began a recall movement to oust the local mayor for *not* proceeding with the project. Hokkaido Electric became concerned that this process, like in the Ashihama dispute, might lead to opposition. It responded by paying 200 million yen to compensate the town for the loss of benefits expected from the site change.[43] The Kyowa community's response was a clear case of the "reverse NIMBY syndrome."

In June 1981, the utility commenced negotiations with the Tomari cooperative. The choice of Tomari reflected a continuation of an earlier strategy relating to the ordering of negotiations between fishing cooperatives and the structure of compensation payments. After the site change, the utility was only legally required to acquire property rights from the Tomari cooperative. Politically, it was still required to offer cooperation money (fishing development compensation) to other cooperatives. It decided to negotiate initially with the Tomari cooperative to improve its position over the other cooperatives with which it would negotiate.[44]

The company negotiated with the Tomari cooperative between 10 and 18 June 1981. The negotiations centered on the relative share of compensation for fishing rights and fishing development. The utility initially offered 180 million yen, comprising ninety million yen for property rights and ninety million yen for fishing development. The cooperative, basing its demands on the Onagawa situation, demanded between 600 and 700 million yen. The utility subsequently offered 220 million yen. In response, the cooperative demanded 500 million yen. On 18 June, the prefectural government intervened and both parties agreed to a 300 million yen figure, comprising 200 million yen for property rights and 100 million yen for fishing development. The utility kept fishing development funds at close to ninety million yen by paying more to purchase fishing rights.[45]

The sequence of negotiations with the other three cooperatives was Sakazuki, Kamoenai, and then Iwanai. As hoped, the utility maintained roughly the same levels of development compensation (seventy to eighty million yen) to all three cooperatives.[46] The Kamoenai cooperative was larger than the others and the prefecture decided to provide a special, one-off subsidy for fishing development there. The completion of these negotiations effectively nullified the impact of resistance by the Iwanai cooperative. Indeed, the utility did not provide cooperation money until after the EPDCC permit was issued in September 1982.

Exogenous events can change expectations in ways that both impede and facilitate bargaining. The terrorist threat hindered siting, but the utility's response effectively eliminated the Iwanai cooperative, its major opponent, from core bargaining processes. A company official lamented, "If you look at the events that followed the terrorist threat, you could come to the conclusion that terrorism helped resolve the conflict. But we could never say that in public, and the company would never be associated with such an interpretation."[47] The company skillfully used its enhanced bargaining position to manage the site change. It negotiated effectively with the Tomari cooperative to provide a low benchmark for subsequent negotiations over cooperation money. It also provided compensation to Kyowa town, which was no longer involved in core bargaining processes. Including Kyowa in the payout was a function of political interest, not good will. The company compensated earlier supporters in Kyowa town to prevent them from opposing and injecting more uncertainties into bargaining.

CONCLUSIONS

In resolving the Tomari dispute, promoters provided one of the largest compensation packages in the history of sited power plants in Japan. The package included fishing compensation (3,050 million yen), community compensation for such things as preferential employment and purchases and some collective goods (4,210 million yen), substantial risk mitigation through a site location change, and abandonment compensation (200 million yen). As at Matsushima, the national community also paid compensation (6,140 million yen) under the Three Laws to the regional community. Although compensation was higher than at Matsushima (if one accounts for the site change from the Kyowa-Tomari location to Tomari), it took the utility nearly four times as long to get community agreement.

The structure of the bargaining environment was fashioned in a way that, according to the earlier quantitative analysis, would have led to a much longer settlement time (131 months) than the average of all of the initial nuclear plant sitings in Japan (eighty-two months). Electricity shortages in both the national and utility spheres were generally low, and there was not a strong market justification for the project. Incomes in Hokkaido were growing at a slow rate, and there was no strong development push for the project. In contrast, the ability of local governments to provide public goods was low, which was consistent with strong local government support. Opportunities in the rural sector explained opposition from this sector. Relatively strong Leftist influence in the prefectural assembly was important in the emergence of ideological opposition. From the late 1960s and early 1970s, communities started to place a higher priority on preservation of the environment.

The quantitative results provide a good explanation of the structure of the negotiating environment. Electricity policy makers generally did not place a high value on the project as a means of meeting electricity shortages. Although there was strong support for the project at the local level, the lack of prefectural support, rural and ideological opposition, and attitudes toward environment preservation dominated bargaining processes. Cumulatively, these conclusions suggest a bargaining environment in which little economic surplus existed and promoters could not easily compensate community interests. All of these factors explain Hokkaido Electric's negotiating approach. It was continually willing to delay the project, even when changed circumstances weakened opposition toward the project.

The magnitude of the residual (twenty-five months) was similar to the Matsushima case, but the model underestimated the time required to resolve the dispute. We need to examine the additional constraints promoters faced in compensating community interests. I suggest that rising project costs were important at crucial stages of the dispute. The costs were conditioned by the relatively small size of the grid, the unusual project design, and the management difficulties operating the project. At several times during the dispute, when the utility was positing itself to negotiate, construction, management, and compensation costs increased quite dramatically, thereby delaying negotiations.

The major costs like at Hamaoka and Matsushima, were spread relatively unevenly across and within regional fishing groups. Only the Iwanai cooperative maintained quite strong resistance—others actually supported the project. Even within the Iwanai cooperative, the costs and benefits were disproportionate across coastal and deep-sea fishing interests, which prevented the industry from sustaining collective action against the project. In contrast, the costs to the prefecture and the utility were quite high for long periods during the dispute but for different reasons. The prefecture saw high costs in terms of coal policy, whereas the utility had grave doubts about the project's economic viability for much of the dispute. Even though the strength of resistance was more like that at Hamaoka, a settlement at Tomari took longer to negotiate because of the utility's unwillingness to provide adequate compensation.

The allocation of bargaining power did not strongly favor the utility for much of the dispute. It had few siting alternatives like the EPDC did at Matsushima. Although it did have alternatives for developing non-nuclear projects, it was more or less locked into the Tomari site for a nuclear plant. This lack of options placed the utility in a weak bargaining position, which reflected the amount of compensation required to win agreement where there was considerable regional support. Fishing and other community interests used their bargaining positions effectively, even when the utility changed the site location.

As in other siting cases, opponents used a range of strategies to enhance their bargaining positions. These included using protest as a political resource to increase the bargaining costs to the promoter by forming an alliance and sensitizing nuclear risks. For example, fishing cooperatives that did not actively oppose the plant cleverly extracted very large compensation payouts. Even though the utility's site selection strategy left a lot to be desired, it used quite innovative strategies during the dispute. These strategies included project rescheduling tactics, risk minimization, and other negotiating tactics to weaken Iwanai fishing resistance. At the local level, mediation by the Iwanai mayor kept bargaining alive in the face of growing local resistance. What stands out in contrast to other siting conflicts was the difficulty in getting prefectural mediation in dealing with fishing and other community groups until late in the dispute.

Expectations changed dramatically and in different directions during the dispute, sometimes bringing the parties closer together and sometimes pulling them further apart. Exogenous changes were most important. Revised expectations after the oil crisis broke this negotiating deadlock but, in contrast to Matsushima, did not resolve the conflict. The terrorist threat shocked promoters. Yet it was the response to this threat that ultimately allowed the utility to reach agreement. The site change effectively weakened the bargaining power of Iwanai fishing interests and reduced significantly the negotiating costs.

Increasingly high levels of uncertainty about the economic viability of the project characterized much of the dispute until a different site was selected. These uncertainties were attributable to the size of the plant in relation to the expected size of the grid in Hokkaido, the costs associated with compensating adequately Iwanai fishing interests, the lack of effective third-party mediation, and the management costs associated with the initial project design. The utility managed this uncertainty in two important ways. First, it rescheduled the project and gave priority to developing an alternative project with less uncertain bargaining and construction costs. Second, it capitalized on the site change, by using compensation and other strategies to isolate Iwanai fishing interests, removing one of the biggest uncertainties in the negotiating process. Like at Hamaoka and Matsushima, the utility used a range of compensation strategies to reduce the political uncertainties inherent in negotiations.

C·H·A·P·T·E·R E·I·G·H·T

NIMBY Politics in Japan

Japanese experience in siting energy facilities does not fit into the traditional NIMBY spectrum of difficult and protracted cases. There are large and significant variations in the times necessary to negotiate power plant siting agreements with local communities. Negotiated siting agreements with compensation do happen quite often in Japan. The existence of these settlements encourages qualification of conventional explanations about conflict, power, and compensation and their relationships with social choice outcomes. Comparative analysis, particularly of similar experiences in Europe, reinforces this conclusion.

The keys to understanding NIMBY processes and outcomes are bargaining and compensation. Participants often develop innovative and creative approaches to conflict resolution. Bargaining on a regular basis often moves conflicting parties toward common ground. Bargaining is inherently dynamic, and participants' expectations about the value of striking agreements have to change if bargaining is to be successful. The process allows participants to make compromises over go gains and losses. Compensation facilitates consensual politics by making dispute resolution a positive-sum game. Even though some participants may gain more than others, bargaining and compensation are more conducive to settlement than alternative approaches that stress using state power and preemption.

The question is not why Japan's consensual politics implies slow conflict resolution but why it yields negotiated deals slowly in some cases and quickly in others. The answer lies in the importance of the structure of the bargaining environment and the politics of compensation. A key conclusion is that delays in negotiating siting agreements are likely to be shorter when the expected net gains from projects are high and when bargaining environments are structured and used by negotiators in ways that facilitate the use of compensation in politically acceptable ways.

THE KEY ISSUES

The Japanese environmental conflict and the facility siting (focused predominantly on U.S. experience) literatures often portray local communities expecting to lose from noxious projects and opposing their development, local communities having the institutional political power to resist projects that they perceive as not being in their interests, and compensation mechanisms rarely working effectively in conflict resolution. Japanese energy siting experience suggests refining these portrayals.

Conflict

Conflict occurs in all siting processes in Japan. It arises not only because of spillover effects but also because communities perceive that they are expected to bear the brunt of the negative costs of projects. In all of the cases described in this book, utility proposals generated resistance. Property rights holders, particularly fishing cooperatives, figured predominantly in local opposition. Local government opposition was important at Ashihama and Tomari because those governments anticipated large economic and political costs from accepting projects. Perceptions about negative spillover effects play an important role in explaining why community interests oppose energy projects.

Conflict also arises within and across other interests at different jurisdictional and market levels because of an unequal distribution of costs and benefits. Local governments were in conflict with other local and neighboring community interests in all of the cases we examined. Local communities were in conflict with prefectural governments at Tomari and Ashihama. Utilities were in conflict with other developers particularly at Matsushima but also at Ashihama and Tomari. National bureaucratic conflict occurred at Matsushima. Understanding siting conflicts often requires explaining why various interests other than developers and "local" interests believe that they will lose while others will gain.

Conflict is not always of the same intensity. The extent to which different interests view noxious facilities as a bad or good bargain and oppose or support the projects is not constant.[1] It varies widely, which is influenced heavily by the social value that different groups attach to plants. At Ashihama, the utility was in an intense conflictual relationship with local community actors because locals perceived the project to be a terrible deal. In contrast, at Hamaoka, there was much more consensus because local community interests saw the project as offering a reasonable deal. The mixture of consensus and conflict varies widely in siting controversies.

Revisionists are right in arguing that the notion of Japan as essentially harmonious has never reflected reality, but they may have vacilated between extremes. Social and political conflict in Japan and elsewhere is not homogeneous in its intensity. Conflict is a relative, not absolute, term and all social choice processes involve conflictual as well as consensual elements. Conflict will dominate some social processes and consensus will dominate others, even in a country as culturally homogeneous as Japan.

Although differences in the intensity of conflict provide an important explanation of the variations in bargaining times, the relationships are often surprising. There was clearly less local community resistance at Matsushima relative to Hamaoka, yet it took developers longer to resolve the former dispute than the latter one. At some early stages of the Tomari dispute, there was virtually no conflict. There were times when both the developer and the fishing cooperatives were ambivalent about the project, which was one important source of delay. The apparent lack of conflict does not necessarily mean consensus. Conflict does not always filter through political processes in consistent ways and generate uniform bargaining results. There is much room for the use of power and leadership in explaining siting outcomes in Japan. Politics matters in bargaining and compensation processes.

These observations are relevant comparatively. Resistance to, and conflict over, siting hazardous facilities occurs in all democratic countries. Although the U.S. literature often suggests that conflict in U.S. society has led to a situation in which it is extremely difficult to site noxious facilities, we observe that Japan and nations in Europe and elsewhere have managed effectively a number of highly intense siting conflicts. We must expect resistance to projects that communities do not want, but we should not assume that conflict leads to political paralysis in siting them.

Power

An important conclusion is that the state, particularly the Ministry of International Trade and Industry (MITI), rarely becomes involved directly in siting disputes. Although it tried to use direct pressure on fishing cooperatives at Ashihama, it failed and has not repeated the performance in energy siting. Even though MITI continues to administer the Three Laws, negotiations between prefectural and local governments determine the allocation of funds. Although the state also grants final approval for siting, it does so only after local mayors and prefectural governors judge politically that property rights transfer and other community bargaining processes are basically finalized. The state is more of an indirect than a direct player in regional siting processes.

MITI's approach has been to delegate the task of siting power plants to private utilities, in contrast to other nations, where public-owned utilities are responsible for siting.[2] Part of this strategy relates to Japanese utilities wanting control over energy markets, but it also relates to local resistance to state intervention in local decision-making processes. Managing NIMBY disputes can be messy, and MITI has set up the necessary incentive and informational structures to facilitate bargaining within and across implementing organizations (utilities) and target groups (community interests). MITI supports utilities in their siting struggles, but utilities and subnational actors do the negotiating. The state uses indirect and implicit, rather than direct and explicit, pressure, as some observers suggest, to influence regional siting processes.[3]

In contrast, local and regional political and economic interests use direct political pressure to resolve NIMBY disputes. Fishing cooperatives exerted considerable pressure on utilities at Ashihama, Matsushima, and Tomari. Deep-sea fisherman wielded power over coastal fishermen to extract large shares of compensation in most cases. Local mediators used pressure to oust a leader of a fishing cooperative at Hamaoka and on individuals at Matsushima. Utilities exerted pressure on fishing cooperatives and other community interests in all of the disputes through alleged bribery, isolating resistance, strategic scheduling of negotiations, and beginning construction.

The ability of different interests to influence siting outcomes varies widely. The state, even with its political gunboat approach, had very little positive influence at Ashihama, compared with its influence at Tomari and Matsushima. The prefecture was more effective in facilitating a resolution at Hamaoka than it was at Tomari and Ashihama. Fishing cooperatives were in much stronger positions at Ashihama and Tomari than they were at Hamaoka. Private companies clearly had more influence at Hamaoka than they had at Ashihama and Tomari. We cannot assume stability across space or over time in power relationships between participants involved in bargaining processes.

Most of the Japanese-related political science literature generally casts the allocation of power in terms of institutional power: state versus local communities, private companies versus local government, or local elites versus local public. The strong-state hypothesis implies that the state is a heavyweight in bargaining processes. It does not mean that the state rules by decree or fiat. The weak-state proposition implies other societal interests, such as local community interests and private companies, are more influential. The Japanese environmental conflict literature concludes that local interests dominate siting outcomes and supports the thrust of the weak-state proposition.

Just as Richard Samuels reminds us to distinguish between *jurisdiction* and *power,* this book suggests the need to delineate *institutional power* from *bargaining power.*[4] Institutional power derived from property rights ownership, protest, dependence on state finance, or monopoly status cannot explain variations in bargaining times. Rather, it is the exercise of bargaining power, defined as the best alternative to a negotiated solution, that is crucial in understanding the observed diversity.[5] Possession or acquisition of institutional power is a ticket to gain access to bargaining processes, but it is relative bargaining power that determines the extent to which competing interests influence siting outcomes.

Comparative examination reinforces these conclusions. Japan has developed energy projects quite well compared with Canada and remarkably well compared with the United States despite strong local autonomy in both of those nations. Japan's performance has been more like France's even though there is arguably less local autonomy and a stronger, more centralized state in the latter. France and other European nations often use economic incentives in siting negotiations. Understanding social choice outcomes requires attention to the structure of bargaining power, the structure of bargaining relationships, and the incentive structures that facilitate negotiations, rather than notions of institutionalized strength or weakness.

Compensation

Compensation aims to redistribute some of the expected benefits of projects from gainers to losers to increase support or at least reduce opposition to projects.[6] Japan has a sophisticated set of mechanisms that it has used historically to compensate property rights holders and other community groups for environmental spillover effects. These mechanisms have given utility negotiators and local leaders considerable scope and flexibility in devising creative solutions to resolve siting conflicts. As the case studies highlighted, Japanese siting negotiators have used almost every compensation instrument.

As Ellis Krauss, Thomas Rohlen, and Patricia Steinhoff conclude, compensation is more generally important in bargaining and conflict resolution in Japan.[7] Compensation works better, however, in facilitating bargaining in some cases than in others. At Ashihama, huge amounts of compensation did not facilitate a negotiated solution. After the site change, compensation did not work either despite strong community support for the project. At

Hamaoka, relatively small amounts of compensation worked very well in striking a deal despite considerable initial resistance. At Matsushima, it worked quite well, despite some bargaining hiccups, although the state had to compensate handsomely both private and fishing cooperatives. At Iwanai, compensation was eventually effective in winning consent after a prolonged dispute, but the utility had to pay significantly more than it wanted to even though a site change weakened key resistance.

Much of the literature about conflict resolution in Japan stresses noneconomic instruments, such as the use of third parties and keeping conflict localized on the original parties, but the literature also recognizes the importance of compensation. Evidence from energy siting also emphasizes these noneconomic strategies in relation to how they facilitate or impede the use of compensation. The cases suggest that bargaining strategies, such as isolating protest, bringing that resistance into decision processes, and design modification, accelerate negotiating agreements only insofar as they act as catalysts for managing compensation requirements. This limited effectiveness points to the importance of interactions between noneconomic and economic mechanisms in understanding siting conflicts.

Several scholars have considered the conditions in which compensation resolves conflict. Susan Pharr implies that "pre-emptive concessions" are likely to head off subsequent conflicts in the same area.[8] Subsequent projects are, however, easier to site not simply because compensation had warded off conflict but because of utilities' property rights acquisition strategies, the economic impacts of projects on local communities, and legal aspects compensation arrangements. Kent Calder suggests compensation is more likely to work in "crisis" situations.[9] Although there was some evidence to support this proposition (settlement times tended to be shorter when utilities expected large electricity shortages), bargaining times actually increased during his defined crisis periods. Howard Kunreuther and Douglas Easterling posit that compensation is likely to work only when some "threshold level of safety" is assured,[10] yet Japan continues to use compensation effectively in siting power plants even after the Three Mile Island and Chernobyl accidents.

We have to go beyond notions of preemptive concessions, crisis, and safety in understanding the relationship between compensation and the resolution of energy siting conflicts. It is the nature of compensation mechanisms, the structure of bargaining environments, and negotiating skills that determine whether compensation is likely to work as an effective policy instrument. Compensation arrangements need to be in place; the net expected gains from projects need to be sufficient to allow developers to compensate losers; and bargaining strategies need to adequately address the politics of bargaining, including legitimacy, interest group behavior, party competition, and distributional issues in dynamic, uncertain, and risky environments.

Japan is more like nations in Europe than like the United States in using compensation to manage siting conflicts. Much of the U.S. literature argues that compensation rarely works in resolving siting conflicts. In contrast, many European countries use compensation (although the nature of their compensation mechanisms differs), which appears to be important in understanding why their siting records are much better than those in the United States. Compensation often works in the resolution of political conflicts. Compensation as a political resource can be used to manage protest as a political resource.

Compensation is not simply calculating costs and buying off resistance. It is also about reducing costs through risk mitigation strategies. Japanese utilities make extensive use of risk mitigation to reduce the overall explicit monetary requirements necessary to strike siting deals. Contrary to Kent Portney's argument, power plant risks are compensatable—community interests do make risk-benefit trade-offs, but the legitimate use of compensation strategies will also determine the willingness of communities to make those trade-offs.[11]

Two structural aspects of compensation are noteworthy cross-nationally. First, Japan's arrangements, like those in France but unlike those in the United States, legally bind property rights holders to negotiated arrangements. This reduces uncertainties associated with possible abrogation of agreements by either side. Second, Japan's arrangements, again like those in France but unlike those in the United States, assume that the effects of noxious facilities are not confined to the administrative units where they are located. In all of the cases examined in this book, interest groups with veto power from adjacent communities were heavily involved in negotiating processes. Comparative analysis suggests that these institutional aspects are important in determining the effective use of compensation as a dispute resolution device.

Japan's NIMBY politics is about how and the extent to which dynamic bargaining moves competing interests toward resolving conflict. It involves understanding that private actions impose an array of costs and benefits on different regional interests and affect the intensity of political conflict. It is about ascertaining the structure of the bargaining environments in which consensus takes place and how the allocation of bargaining power allows some actors to influence siting processes more than others. It is concerned with the importance of redistributive schemes and the ability to manage creatively compensation requirements.

THE PROPOSITIONS AND THE EVIDENCE

I hypothesize that the ease of negotiating siting agreements is related to the structure of the bargaining environment, the distribution of costs and benefits, bargaining strengths, negotiating skills, changing expectations, and uncertainty. I summarize the major findings of this book and then offer some historical perspectives on Japan's energy siting conflicts. I conclude by relating my conclusions to broader Japanese studies and comparative siting literatures.

Polimetric Insights

The structure of the bargaining environment is the most important factor influencing negotiating times in Japan, explaining a large proportion (75 percent) of the variations in those times. The value that participants place on expected costs and benefits from projects conditions their responses to those projects. This value influences the aggregate surplus that is available for redistribution from gainers to losers and determines whether the overall bargaining environment acts as a positive or negative catalyst for negotiating social agreements. The examination suggests considerable predictability in the nature of participants who become involved in siting disputes, the patterns of those participants' responses to projects, and the relationships of those responses with approval times.

The responses of developers are important factors influencing bargaining times. It is easier to site projects when developers expect larger shortages because they are more willing to compensate community interests, although this reaction is stronger in reducing approval times for initial packages. Bargaining delays are not always caused by local community opposition. Even when little opposition exists or support is relatively strong, developers do delay projects intentionally when projected shortages are low. For example, at Tomari, the utility was simply not interested in negotiating because the economics of the project did not justify developing it quickly. Comparatively, one critical reason for the slower growth in energy capacity in the United States is that electricity growth rates have been much lower in this country than in Japan and some European nations.[12] It would not be economically rational to site power plants even if there were *no* community resistance.

Local and prefectural governments and other community interests have veto power over placement decisions and their responses to offered projects also determine bargaining times. Responses to hazardous facilities are influenced not only by incomes but also by a host of other financial, industrial, political, ideological, and attitudinal factors. Community responses shape the structure of the bargaining environment in which settlements are negotiated. Understanding the strength of political demands for environmental preservation is more complex than the standard environmental quality hypothesis acknowledges.[13]

Negotiations tend to come to fruition more easily when levels of social and economic opportunities are expected to rise rapidly, although this effect is stronger in shortening settlement times for nuclear plants. Under conditions of rapid economic growth, a boom economy mentality usually emerges and there is less concern about environmental degradation. For example, the Mie and Shizuoka prefectural governments, whose economies were growing quite rapidly, were clearly much more interested in energy plants than the Hokkaido government, whose economy was growing more slowly. As this impact is stronger in facilitating nuclear settlements, it suggests that the perceived nuclear risks may be more acceptable to communities that place a stronger emphasis on economic growth.

Resolving conflicts is easier in situations in which the local financial base is weak, although this is more important in shortening approval times for subsequent packages. This effect on conflicts supports the proposition that economic determinants are important in setting regional policy priorities in Japan.[14] It was highlighted in the responses of the Kisei (at least early in the dispute), Hamaoka, Ooseto, and Tomari governments, whose finances were unhealthy. As David Morell and Christopher Magorian point out, "In energy facility siting cases, [local] officials have tended to show a bias in their decision making toward policies that are compatible with the management of the community's budget."[15] In Japan, these structural biases are particularly strong in subsequent packages, where demands for the continued provision and maintenance of public goods are strong because construction-induced, economic booms come to an end after operation commences. Local governments tend to support subsequent projects more readily because not doing so can have significant political and electoral costs.

Negotiating deals over risky projects is easier when local Leftist political parties are relatively weak, although this effect is stronger in lengthening approval times for nuclear plants. Partisanship is an important variable influencing the ease of resolving environmental conflicts. Leftist opposition was clearly an issue at both Hamaoka and Tomari, although at Ashihama it was conservative opposition that forced termination of the project. This effect qualifies the proposition, put forward most eloquently by Samuels, that local partisanship is a "red herring" and "does not relate to policy [positions] or extra-local relations."[16] Studies in the United States also suggest a relationship between partisan identification and attitudes toward the environment, although French experience seems to suggest otherwise.[17] In Japan, ideology matters, particularly when Leftist parties have policies that oppose certain categories of projects. Ideological values are less easily compensated than economic ones in the siting of nuclear projects.

Social consensus is more easily achieved when social and economic opportunities in the rural sector are relatively scarce, although this effect is stronger in reducing settlement times for conventional and initial packages. The ability of rural property rights holders to influence bargaining varies greatly. Fishing cooperatives at Ashihama opposed vehemently because of an economic boom in the industry. In contrast, some fishing cooperatives at Hamaoka, Matsushima, and Tomari actually supported, or at least did not actively oppose, energy facilities because they perceived that compensation would alleviate financial problems. It is clearly more difficult to provide adequate compensation to rich property rights owners. In comparison, opposition by fishing interests in the United States forced planners to avoid many coastal locations and instead use inland sites with cooling towers.[18] European wine growers have been notoriously anti-nuclear.[19]

Hazardous facilities are located more easily where social attitudes place low emphasis on environmental preservation, although this effect is stronger in shortening bargaining times for subsequent packages. This fact provides qualified support to a dominant conclusion in the literature that siting became more difficult during the 1970s with the strengthening of citizens movements throughout Japan. This conclusion also helps explain why Japan and some countries in Europe have been able to continue siting subsequent power plants during the 1980s and 1990s.

The quantitative evidence sheds light on puzzles about the interrelationships among conflict, bargaining, and compensation. As the Tomari and Hamaoka cases showed, similar levels of initial resistance to unwanted projects can lead to very different siting outcomes. At Tomari, the developer was not willing to provide adequate compensation to negotiate a deal to develop an inefficient project. At Hamaoka, the developer supported the project and was willing to provide adequate compensation to community interests. Bargaining environments that are conducive to proponents compensating losers are likely to yield faster siting agreements.

Qualitative Insights

Although this analysis could end after generating polimetric models (and many quantitative social scientists would do so), there is much to be gained by enriching the analysis with case studies. Other nonstructural factors, such as the distribution of costs and benefits, leadership and bargaining skills, changing expectations, and uncertainty, also influence the course of siting conflicts. Quantitative models provide a way of identifying disputes when bargaining structures are dominant and ones when they are less so in determining siting outcomes. Unfortunately, statistical techniques offer little to aid in understanding the nature of bargaining and compensation processes and how bargaining environments interact with other factors and affect the management of siting conflicts.

Negotiating siting deals is easier where the distribution of costs is spread more widely across community groupings because the ability to engage in collective action is weakened. The Ashihama case showed that fishing cooperatives were able to sustain resistance because the per-capita costs of the project were heavily concentrated on these pre-organized groupings in a regional economy characterized by productive specialization. Other cases demonstrated that fishing cooperatives were unable to engage in effective opposition because the costs of projects were spread more unevenly across interests operating in more diversified economies. Although that resistance was generally quite fierce initially, opponents could not overly delay bargaining processes.

Collective action opposing noxious facilities is a feature of politics in all industrialized nations. The extent to which such resistance impedes bargaining and the use of compensation mechanisms depends on the magnitude of the per-capita costs of projects and the degree to which those costs are spread evenly across losers. When there are high per-capita benefits of engaging in collective action, groups will have an organizational advantage because they will be more willing to mobilize resources to achieve their objectives. Stronger and sustained resistance is a feature of some siting processes and not others, even though overall costs of projects are generally concentrated at the local level.

Siting agreements are more forthcoming where communities have less bargaining power than proponents. At Ashihama, local opponents had far more bargaining power because they had excellent alternatives to a negotiated solution. At Hamaoka and Matsushima, bargains were struck fairly quickly because community interests had very few viable options to accepting projects. At Tomari, conflict resolution was delayed because bargaining power was more or less equally distributed among parties—for much of the dispute both the developer and the key opposing fishing cooperative had viable alternatives to negotiating a deal.

The negotiating power of different state, regional, and private sector interests is conditioned by the extent to which they have alternative courses of action available to them. These other options set not only the limits of bargaining but also the extent to which parties can use bargaining power to negotiate more or fewer compensation requirements or offers without making alternatives more attractive to themselves or others. Some groups, which seemingly have considerable institutional power, cannot influence some bargaining processes very much at all. Going beyond notions of the strong-state–weak-state distinction, no matter how sophisticated they may be, and exploring politics in terms of bargaining power can clarify which groups have power and under which conditions and terms in Japanese society.

Conflict management is easier when negotiators execute strategies that bring conflicting parties to the negotiating table in ways that participants believe to be politically legitimate. At Ashihama, developers clearly lacked skills to deal with intra-party conflict, and opponents were very skillful in forming an alliance and gaining electoral access to power. At Hamaoka, promoter skills were crucial in preventing conservative intra-party conflict from becoming an issue, in managing concerns about risk, and in splitting an opposition alliance. At Matsushima, proponents were rather skillful in handling commercial and bureaucratic resistance although they did not manage local bargaining in a way that kept their earlier promises. At Tomari, promoters were adroit in the way they delayed negotiations to keep open the option of regrouping at a later time when circumstances permitted.

In most cases, opposition groups form alliances to improve their respective bargaining positions.[20] The extent to which those coalitions are "temporary marriages" depends heavily on the spread of costs across members.[21] Pharr suggests a key negotiating strategy of marginalizing resistance, whereas Calder suggests one of bringing that resistance into decision-making processes.[22] Skill in negotiating siting deals requires both isolating resistance that does not matter or does not matter very much and bringing core resistance in bargaining processes so that promoters can influence it. Core resistance has to be dealt with ultimately, either by negotiating with it or by abandoning those bargaining processes.

An important conclusion is that leadership skills in executing bargaining strategies differ quite widely. Some literature suggests that leadership skills do matter in Japanese politics, but we are still confronted with conclusions that Japan is leaderless. A subnational perspective on siting reveals that Japanese leaders are not faceless, robot-like machines who have fixed solutions to all similar negotiating problems. They are often creative, sensitive, and skillful in managing complex bargaining. As some observers recognize, Japan has had some "great" leaders at the national level.[23] But it also has had some "great" leaders and negotiators at the subnational level.

Striking social agreements tends to be easier when changed expectations, either exogenously and endogenously induced, increase the net gains expected from projects. At Ashihama, endogenously induced changes in expectations through use of political pressure and inappropriate compensation strategies created more resistance. At Hamaoka and particularly at Matsushima, exogenously induced changes in expectations through both domestic and international energy market changes were critical in allowing proponents to offer compensation to nullify resistance. At Tomari, exogenously induced changes in expectations both facilitated and impeded bargaining, although it was ultimately a combination of exogenous change (terrorist threat) and endogenous change (plant relocation) that allowed the utility to strike a deal.

Bargaining is inherently dynamic, and static models of politics will not capture processes and outcomes effectively. If bargaining is to lead to consensus, someone's expectations about the relative value of striking a deal must change or conflict will remain deadlocked. But it is the direction and magnitude of changed expectations that is important. Fundamental changes that increase dramatically the expected costs of bargaining may stall consensus formation. Smaller changes that increase marginally the expected benefits of negotiating may not speed up consensus. The fact that many disputes are resolved quite quickly suggests that expectations, for whatever reason, do change, and sometimes very rapidly and fundamentally. In turn, changes in expectations lead to changes in the structure of bargaining relationships and compensation offers and requirements.

Negotiations tend to yield faster agreements when there is less uncertainty about the value of those agreements. At Ashihama, negotiators abandoned bargaining at two sites because they were highly uncertain as to when they could resume local bargaining and at what cost. At Hamaoka, large net expected benefits allowed negotiators to deal with uncertainty very effectively, particularly in symbolic terms. At Matsushima, high levels of uncertainty in dealing with resistance turned into more certainty (or at least provided more benefits to rectify that uncertainty) when the oil crisis occurred. At Tomari, consistently high levels of uncertainty about the economics of the project led negotiators to postpone bargaining over the project and to give priority to other, less uncertain projects.

Uncertainty is a common feature of all complex bargaining involving multiple actors. It is particularly important when negotiating stakes are high. We cannot assume that participants have good knowledge about the costs and benefits of bargaining, their ability to use power to influence bargaining, and their effectiveness in using compensation mechanisms. Uncertainty is a key factor in explaining why conflict emerges and why some controversies are resolved in ways different from others. Those involved in bargaining know about and appreciate the variable nature of uncertainty. Consensual politics is about managing uncertainty around the complex and often interrelated compromises that must be struck.

Siting delays are not fixed even after accounting for the structure of the bargaining environment. The Ashihama project was abandoned because promoters, despite offering large amounts of compensation, could not manage the politics of intra-party conflict effectively. The Hamaoka project was developed more quickly and with little compensation because proponents managed intra- and inter-community distributional issues quite effectively. The Matsushima project was also developed quite quickly but with very high compensation because changed expectations allowed developers to skillfully compensate opposing interests. Explaining siting delays requires exploring interactions between bargaining structures and negotiating skills.

Historical Perspectives and Contemporary Developments

The dominant story on locating noxious projects in Japan suggests that siting became more difficult in the late 1960s and early 1970s with the advent of citizens movements that used "protest as a political resource" in Japanese politics.[24] The siting of energy facilities clearly got caught up in the emergence of these movements. The average time it took developers to obtain community agreements for energy projects has generally increased since the late 1960s as developers face increasing levels of environment protest.

The evidence from energy siting suggests two important qualifications. First, power plant siting in Japan has been essentially a two-edged account of fishing cooperatives and compensation. During the early twentieth century, Japan established fishing cooperatives to manage common pool resource problems associated with inland fisheries. Fishing cooperatives have been central in protesting against energy facilities. During the 1960s and 1970s, Japan built on earlier redistributive schemes and established specific compensation mechanisms for ameliorating opposition by fishing cooperatives and other local community interests to the siting of energy projects. This account is essential in understanding opposition to energy projects and the management of environmental conflict.

Second, many proposed nuclear plants during the 1960s actually took very long times to site or were abandoned, even though there was a high priority on economic growth and less concern for the environment. Furthermore, settlement times for approved nuclear plants have declined since the mid-1980s. The reason lies in distinguishing between different types of power plants. Since 1983, utilities have not sited any initial nuclear plants; they have been able to develop only subsequent nuclear plants, which have shorter settlement times. This phenomenon highlights the danger of periodizing siting history without regard to the diversity of times in siting different types of projects.

From the early 1980s to the present, Japan's nuclear capacity expansion has occurred only at initial nuclear sites that were opened between the mid-1960s and the early 1980s. Subsequent packages have been consistently easier to site than initial packages because structural biases provide strong incentives for both developers and communities to negotiate deals for more plants after the first ones are built. Utilities are likely to continue to develop subsequent nuclear projects relatively quickly for some time because of the YIMBY response to those projects.

The real litmus test for Japan's nuclear industry will come when subsequent nuclear sites become exhausted and when any future growth in nuclear capacity becomes dependent on opening initial nuclear sites. It will be at that point (the exact timing will depend on future electricity demand growth) that Japan will face a major crossroads in the development of nuclear power. Utilities are likely to face bargaining times of twenty to thirty years for initial nuclear projects.

Continued opposition to siting the Maki nuclear plant, which Tohoku Electric proposed in 1971, illustrates these difficulties and highlights possible new challenges. The Maki dispute resembles a hybrid between the Tomari and Ashihama disputes. For much of the dispute, the plant was simply un-economic given the small size of the company's grid. Strong resistance by the fishing cooperatives, a Leftist anti-nuclear movement, and intense conflict within the local Liberal Democratic Party (LDP) stalled bargaining processes. The developer's bargaining skills left much to be desired—the utility allegedly used bribery tactics, it overly relied on one LDP faction, and it did not acquire all of the necessary property rights to build the project. The utility and the local leadership were confronted with a recall movement.[25]

But what really worries nuclear developers is that anti-nuclear community groups at Maki pressured local politicians to hold, for the first time in the history of nuclear siting, a local plebiscite in 1997 to decide whether to accept the project. There was a very high turnout rate (90 percent), with an overwhelming 61 percent voting against the project. Although legally the results are not binding on the local community, politically they are significant. Several other local communities have pressured their governments to hold plebiscites, including Kyushu Electric's Kushima initial nuclear plant. The Maki result is likely to strengthen the positions of fishing cooperatives and other anti-nuclear groups at those sites.[26]

Like negotiating times, aggregate compensation requirements for winning community acceptance for initial projects have increased since the 1960s. Siting the Hamaoka plant in the 1960s required about 2,000 million yen and some marginal risk mitigation. Obtaining agreements for the Matsushima project in the 1970s and Tomari project in the 1980s required about 14,000 million yen at each. At Tomari, it also required considerable risk mitigation in the form of a site change. Together with compensation offered in the abandoned Ashihama nuclear case (about 8,000 million yen and a site change), these deals suggest that the cost at which developers are willing to give up particular sites has increased. Compensation requirements have increased about seven-fold.

It has been fishing cooperatives who have been the really strong and clever bargainers in energy siting disputes. Fishing cooperatives have received increasing per-capita outlays for transferring their fishing rights to utilities, from small amounts in the 1960s (Hamaoka: 0.2 million yen) to much larger amounts in the 1970s and 1980s (Matsushima: 2.3 million yen, and Tomari: 2.1 million yen). These agreements suggest a ten-fold increase in per-capita utility outlays to win consent from fishing cooperatives, who have also obtained large proportions of indirect compensation under the Three Laws, even though the Three Laws were set up to provide compensation for non-owners of property rights. Fishing cooperatives have been able to extract increasing compensation from both utilities and the national community in return for their acceptance of energy plants.

Utilities have clearly moved into more costly sites for initial nuclear projects in terms of both bargaining times and compensation requirements. As initial sites becomes increasingly scarce, fishing cooperatives will probably generally be in even stronger bargaining positions with utilities over the opening of new nuclear sites. Given the nature of Japan's compensation mechanisms, it is likely that fishing cooperatives will continue to demand high, or even higher, compensation packages. The recent Maki referendum result may given them an additional bargaining chip.

The 1995 revisions to Japan's Electric Utility Industry Law are likely to create more diversity in the politics of energy facility siting.[27] Against a background of comparatively high electricity prices in Japan and increasing electricity demand, the changes sought to inject more competition into electricity markets by allowing the limited entry of Independent Power Producers (IPP) into those markets to develop conventional energy projects. Many of the IPPs that have successfully tendered to enter the market were companies from declining industries, such as cement and petrochemical industries, that have in-house generating capacity.

These new competitors are crucial in terms of siting because their presence increases the availability of urban sites on which to develop conventional power plants. These IPPs, with much of their existing industrial capacity close to major urban areas, have already possessed the land on which to develop energy projects. More important, they have already acquired fishing rights from fishing cooperatives, which means that IPPs will generally be developing energy projects on sites that can be regarded as subsequent sites, where property rights negotiations are not likely to be problematic. But at the same time these urban sites have increasingly stringent environmental controls.

These developments point to an emerging locational pattern in Japan that is different from the past ruralization pattern. Utilities will continue to be limited to siting new nuclear plants in increasingly rural locations. IPPs will have the opportunity to site many conventional power plants in more urban areas. The two types of utilities suggest a split pattern of continued ruralization of nuclear siting and more "urbanization" of conventional plant location. The dual challenge facing Japan's energy policy is managing rural environmental disputes in siting nuclear plants and dealing with urban environmental conflict in developing conventional power plants.

ANALYTIC AND COMPARATIVE IMPLICATIONS

I conclude by exploring the analytic and comparative implications of this book for Japanese politics and facility siting more generally. The comprehensiveness of the examination encourages rather than discourages such broader generalizations.

Most general analyses of Japanese politics tend to focus on the distribution of power at the national or local levels. The evidence assembled in this book highlights the need to integrate different levels of government and societal interests into analyses of Japanese politics rather than to treat them separately. Many key policy issues in Japan have both national and subnational dimensions. When national policy action or change influences subnational governments, the interests of subnational governments and how they and other local interests respond will influence political outcomes. Bargaining will then take place both within and across different jurisdictional and market levels, suggesting the centrality of bargaining between and within local and regional private, governmental, and other community interests, often with the state playing an indirect role. Subnational governments clearly do matter in Japan's politics and policy processes.

Despite a great deal of literature arguing the contrary, we are still confronted by studies that stress uniformity in Japanese public policies and politics. Japanese energy siting history highlights diversity in the structure of bargaining environments, particularly with respect to the intensity of conflict, power relationships, and the effectiveness of conflict resolution mechanisms, within the same policy arena. Japan's consensual politics yield considerable variety in terms of negotiating solutions and outcomes. Approaches that stress uniformity are useful only when variations in what they are explaining are small and insignificant. When these variations are significant, these approaches need to employ differentiated models to explain diversity in Japanese politics.

The analysis of energy siting processes and outcomes suggests that Japan is not a generic nonresponder and is not leaderless. In siting, negotiators and leaders often respond to changed circumstances and develop innovative approaches and strategies to deal with political siting problems. Responding to what others do and do not do and even inducing responses from others in moving toward consensus is an inherent feature of bargaining. In contrast to their Western counterparts, Japanese leaders and negotiators may be less charismatic and often work privately behind the scenes.[28] But they do develop sophisticated strategies for dealing with highly complex conflictual situations—a skill that can lead to different bargaining solutions and outcomes. Consensual politics is about diversity in the management of conflict in creative and innovative ways and learning from past experiences, whether positive or negative. This definition of consensual politics emphasizes the importance of not only differentiated but also creatively differentiated models of Japanese politics.

Japan's siting experience also sheds light on project siting in theoretical and comparative contexts. Most standard site-selection approaches use least-cost methods, which rank projects in terms of economic efficiency and identify high-return projects. These approaches ignore the importance of politics of conflict, bargaining, and compensation. They assume the social and political costs involved in winning approval are not relevant to benefit-cost calculations because some redistributive agency is looking after the politics of distribution. In short, they presume zero implementation costs in resolving siting conflicts. Policy makers will however be interested in political implementation costs as this will influence their ability to achieve policy objectives that require the development of those projects. Bargaining and compensation costs need to be taken into account at the evaluation stage.

This strategy provides a way for developers, both public and private, to consider more effectively the sources of delay in resolving NIMBY conflicts. The case studies suggested that utilities had major difficulties in gauging settlement times. The development of models that can generate reasonable estimates of approval times could be used as an initial screening device for the selection of "low political resistance" sites from a pool of "least economic cost" sites. Other nonstatistical techniques could assist further investigations about the composition of interest groups, patterns of property rights ownership, likely strategies in relation to local political issues, and ways to manage compensation requirements. These more detailed analyses would best be done once the number of candidate sites was narrowed sufficiently. Such an approach would provide more information about expected implementation times and could assist in reducing the high levels of uncertainty that developers confront in siting.

The importance of bargaining and compensation, as Joseph Cordes and

Burton Weisbrod point out, has particular import for this book: "Economists have produced a substantial literature on the theory of compensation for persons harmed by public actions; but until recently there has been relatively little research on the institutional and political aspects. However, the political attractiveness of any public policy depends critically on how its costs and benefits are distributed, an aspect that can be affected greatly by the provisions of compensation."[29] I concur fully with this statement. Those involved in making siting decisions often do not have control over implementation. Siting involves bargaining between relatively autonomous and conflicting groups over the distribution of expected gains and losses arising from projects.[30] When there is a lack of power or a lack of will from the top to impose those losses on specific groups, compensation assumes a role as a crucial policy mechanism for redistributing gains to losers and resolving conflict.

Collectively, the polimetric and case study results account for a large proportion of the variation in bargaining times in both urban and rural Japan without recourse to sociocultural aspects of those processes. This conclusion is significant because siting has increasingly taken place in rural areas where we might expect traditional cultural mores and beliefs to affect bargaining processes the most. It suggests that there is a set of structural and institutional (as opposed to idiosyncratic) variables that are likely to determine siting outcomes more broadly. The coefficients would probably not be exactly replicated in other siting disputes whether in Japan or elsewhere. There may be different institutions, interests, decision-making procedures, and definitions of social and economic statistics; however, the evidence points to a set of more general variables that are likely to be useful in examining siting conflicts comparatively.

Japan and European nations that have more institutionalized mechanisms to compensate for environmental spillover effects are doing much better in the siting of energy facilities than countries such as the United States, which does not. Constitutionally, the Japanese state, like many others, has eminent domain powers that in principle it can use to forcibly resume property to site facilities. Japan has adopted, however, and for good reason, an alternative approach that emphasizes bargaining and compensation, in contrast to several states in the United States that have in the past attempted to bypass local opposition.[31] It is clear from relative siting performances that the use of preemption is less effective than bargaining and compensation in resolving siting conflicts, although the mere existence of compensation schemes does not guarantee stability in siting outcomes.

We need more comparative analyses to build better theories about NIMBY processes and outcomes, but we need to exercise caution in identifying relevant comparative reference points. All too often, the Japanese studies literature, even when it is comparative, chooses the United States as a comparative locator.[32] This selection obviously reflects the weight of U.S. scholarship in the Japan studies field. Such a comparison for energy siting reveals a picture of diverse bargaining outcomes in Japan and relatively bad ones in the United States. Expanding our examination further yields a fundamentally different picture.[33] It is U.S. siting experience that is different comparatively. Japan looks more like Europe in terms of bargaining and the use of compensation mechanisms.

Pioneering research by Michael O'Hare, a leading social choice expert, proposed bargaining and compensation as a solution to U.S. NIMBY problems.[34] What is striking is that Japan has been practicing what he suggested for over a century, and Japan is not alone in this regard.

Notes

1. Conflict, Bargaining, and Compensation

1. Popper 1983.
2. Whereas the Narita airport took a long time to site (Apter and Sawa 1984), the Kansai airport and other regional airports took much less time. The new bridge linking Kyushu and Shikoku was developed more quickly than the Tokyo bridge. Many petrochemical projects located close to Tokyo were sited much more easily than the Mishima case.
3. See McKean 1981; Lewis 1980; Samuels 1983; Broadbent 1986; Krauss and Simcock 1980; and Apter and Sawa 1984.
4. See Lesbirel 1988.
5. Samuels 1987.
6. The same conclusion also applies to a range of other energy policy issues (Lesbirel 1994).
7. The best literature on conflict includes Najita and Koschmann, eds. 1982; Pempel 1982; Samuels 1983; Krauss, Rohlen, and Steinhoff, eds. 1984; Mouer and Sugimoto 1986; Samuels 1987; Calder 1988a, b; Eisenstadt and Ben-Ari 1990; and Pharr 1990.
8. Pempel 1982:3.
9. Samuels 1987:20.
10. Pharr 1990.
11. Gamson 1975 provides the analytic underpinnings of these analyses.
12. See Lipsky 1968:1144.
13. McKean 1976; McKean 1981, and Samuels 1983.
14. Singer 1980:58. On the theory of collective action, see Olson 1965. O'Hare 1977; Morell and Magorian 1982; and O'Hare, Sanderson, and Bacow 1983 have related Olson's model to hazardous facility siting.
15. O'Hare 1977. Some interests in these constituencies may gain more than others.

16. See, for example, McKean 1981; and Morell and Magorian 1982.
17. See Nelkin 1977 (Austria, Sweden, and The Netherlands); Nelkin and Pollack 1981 (France and West Germany); Touraine et al. 1983 (France); and Rabe 1992 (Canada).
18. Pressman and Wildavsky 1974.
19. For earlier, although still relevant, overviews, see Fukui 1977; and Calder 1988b: intro.
20. Johnson 1982. Williams 1985 provides an excellent review of Johnson's book.
21. Political scientists include Kabashima and Broadbent 1986; Pempel 1987; Samuels 1987; Friedman 1988; Calder 1988b; Okimoto 1989; Rosenbluth 1989; and Ramseyer and Rosenbluth 1993. Economists include Nakatani 1984; Aoki 1988; and Sheard 1991. A very insightful discussion of the strong state–weak state notion can be found in McKean in Allinson and Sone, eds. 1993.
22. Samuels 1987.
23. Inoguchi and Iwai 1987; and Ramseyer and Rosenbluth 1993.
24. Allinson and Sone, eds. 1993.
25. See Muramatsu 1986; Steiner, Krauss, and Flanagan 1980; McKean 1981; and Reed 1982, 1986.
26. See Steiner 1965 on the historical lack of local government autonomy in Japan. See also Reed 1982 on the autonomy of local Japanese publics.
27. Lewis 1980.
28. The preemptive approach argues for placing all decision power in the hands of state or national governments by establishing, for example, "super siting agencies" and giving them preemptive powers over local decisions. See Baram 1976; and Murray and Seneker 1980. For a good critique of the preemptive approach, see Granger and Wise 1980.
29. See Morell and Magorian 1982; Andrews and Pierson 1985; and Ducsik, ed. 1986.
30. Rabe 1994.
31. Pringle and Spigelman 1981.
32. Morell and Magorian 1982.
33. Environmental Planning Agency 1979. Hence, the acronym NIMTOF (not in my term of office).
34. Haley 1987; Samuels 1987; Calder 1988b; Okimoto 1989; and Allinson and Sone, eds. 1993. See also Aoki 1988 for an economic perspective on bargaining in Japan's micro economy.
35. Fisher and Ury (1983: chap. 6) describe this as BANTA (best alternative to a negotiated agreement).
36. On privatization and mediation, see Krauss, Rohlen, and Steinhoff 1984; Upham 1987; and Pharr 1990. On coercion and intimidation, see Mouer and Sugimoto 1986; and Van Wolferen 1989.
37. Donnelly 1977.
38. Krauss, Rohlen, and Steinhoff, eds. 1984.
39. It is important to distinguish between *ex-ante* compensation, for "anticipated" costs for siting noxious facilities, and *ex-post* compensation, for "actual" costs from such things as pollution incidents that have happened at those facilities.

On the use of compensation for the latter, see Huddle and Reich 1975; and Gresser, Fujikura, and Morishima 1981. On the use of compensation in resolving conflicts in other policy arenas, such as agriculture, see Campbell 1977b; and Donnelly 1977.

40. Apter and Sawa 1984.
41. Samuels (1983:153) discusses how the Ministry of Home Affairs compensated Leftist political parties by reintroducing the direct election of ward chiefs to get their support for the Tokyo Bridge.
42. Most countries use compensation mechanisms in the resolution of siting controversies, although their nature and extent appear to differ widely. See OECD 1980.
43. The problem of binding parties to agreements is discussed in Bacow 1980; O'Hare, Sanderson, and Bacow, 1983; and Bacow and Wheeler 1991.
44. Condron and Sipher 1987.
45. Portney 1991.
46. Pharr 1990.
47. Calder 1988b:471.
48. Kunreuther and Easterling 1991; and Kunreuther, Slovic, and MacGregor 1996.
49. See Roberts 1981 for an analysis of strategic skills of utilities.
50. Baram 1976; and Morell and Magorian 1982.
51. Bradford and Feiveson 1976.
52. For excellent discussions on bargaining, see Schelling 1960; Sabatier and Mazmanian 1979, 1980; Raiffa 1982; Fisher and Ury 1983; Barrett and Hill 1984; and Susskind and Cruikshank 1987.
53. Making people just as well off after the change as they were before it is the well-known potential Pareto improvement criterion. See Pearce 1971; and Mishan 1972 for nontechnical descriptions. On compensation and siting, see O'Hare 1977; Bacow and Sanderson 1980; Hadden and Hazelton 1980; McMahon et al. 1982; and O'Hare, Sanderson and Bacow, 1983.
54. O'Hare 1977:414.
55. Hansen 1984.
56. There is a substantial economic literature on the measurement of external costs and benefits. See, for example, Hettich 1969; and Nash, Pearce, and Stanley 1975. The earliest studies to recognize the importance of nonneutral spillover effects and public participation seemed to be by geographers: Austin, Smith, and Wolpert 1970; and Mumphrey and Wolpert 1973. See also Ervin and Fitch 1979.
57. George 1982:7.
58. Fisher and Ury 1983.
59. The best examples can be found in Fukui 1977; Pempel 1977, 1982; and Calder 1988b.
60. Prestowitz 1988; and Van Wolferen 1989 have had considerable popular impact.
61. Van Wolferen 1989; and Calder 1988a.
62. The most eloquent conceptualization of this notion can be found in Pempel 1982.
63. See Pearce 1971; and Mishan 1972.
64. See IEA 1996.

2. Project Siting and Compensation

1. Samuels 1987:33.
2. Ibid.
3. A comprehensive list of the laws governing the operation of the electric power business in Japan can be found in Shigen enerugii chō 1985.
4. For an excellent although rather technical discussions of how electric power companies formulate electricity capacity plans, see Shin denki jigyō kōza 1980b; and Nihon denki kyōkai 1981.
5. Jopling 1974. For an excellent and still relevant review of the literature, see Hamilton 1979. Refer to Keeney 1980 for the most sophisticated examination of techno-economic aspects of site selection.
6. See OECD 1975; and Toyoda 1976.
7. Nihon genshiryoku sangyō kaigi 1989:13–16. The only exception, apart from hydroelectric power, is a few small, coal-fired power plants in Hokkaido. This exception reflects the existence of inland coal deposits and a preference for minimizing transportation costs. In the United States and other countries, cooling towers are used for construction inland. Japan cannot use cooling towers because inland areas are mountainous.
8. Interviews, 1985.
9. The definition of the stages differs in some cases. See Lester 1983.
10. The reason for only using ninety cases and not the hundred mentioned in Chapter 1 is that the remaining ten were abandoned nuclear plants. Using them would distort these results because each has a infinite lead time. I address abandoned cases in Chapter 3.
11. Indeed, some would argue that complete public acceptance is never reached.
12. Power companies are required by law to provide details of construction plans to MITI on an annual basis; an invitation by a subnational government generally indicates that a power company has discussed the proposal at the local level and has decided to attempt location in that locality. Interviews, 1984.
13. Interviews, 1984.
14. On how these monopoly rights were derived, see Shin denki jigyō kōza 1980a. See also Murota 1984:239–266.
15. Interviews, 1984.
16. The best description of EIAs and public hearings in Japan can be found in Inaba 1977:203–211.
17. Interviews, 1984.
18. See Shakai keizai kokumin kaigi 1981:1–40 for an analysis of licenses and permits required to site power stations.
19. Interviews, 1984.
20. There are two ways of measuring public acceptance times. One is to give all plants in a package an equal public acceptance time. The other is to give the first plant its true public acceptance time and to assign a zero public acceptance time to other plants in the package. The choice of measurement affects average lead times and COVs. In contrast, taking average licensing and construction times for plants in the package does not lead to large differentials in results compared with the unit plant approach and at the same time prevents problems in measuring public acceptance times.

21. Differences will influence measurements more for plants with shorter public acceptance lead times compared with those with longer ones. Although discrepancies prevent precise measurement, they do not alter the overall patterns that emerge from the analysis.
22. Interviews, 1984.
23. See Jopling, Gage, and Schoeman 1973; Knuth and McEwen 1977; Rad 1979; Gladwin 1980; Bacow and Sanderson 1980; Quirk and Terasawa 1981; Morell and Magorian 1982; Reed and Young 1983; Lester 1983; and Radlauer, Bauman, and Chapel 1985.
24. On the legal and ethical dimensions of compensation, see Costonis 1975; and Michelman 1967, respectively.
25. Cordes and Weisbrod 1985 provide an excellent discussion of various types of compensation.
26. On various forms of compensation, see Morell and Magorian 1982:chap. 5. Portney (1991:chap. 7) provides a good analysis of risk substitution. Reich 1991 offers an excellent analysis of symbolic compensation.
27. Tsūshō sangyō shō 1963a,b. These compensation principles are based on *Kōkyō yōchi no shutoku ni tomonau sonshitsu hoshō kijun* (Compensation standards accompanying the acquisition of public lands). See Kobayashi 1983.
28. See Shin denki jigyō kōza 1980b.
29. The following is based on Sato 1978, which offers an analysis of the legal history of the fishing industry.
30. See Mizumoto 1980 for a discussion of land rights in Japan.
31. This discount rate changes according to different interest rates.
32. Interviews, 1984.
33. For a very interesting discussion of this issue, see Kretzmer 1979, and Bacow and Wheeler 1991.
34. Tsūshō sangyō shō 1982; and Shigen enerugii chō 1987.
35. Tsūshō sangyō shō, various issues, 1960–1982.
36. See Krauss and Simcock 1980; and McKean 1981.
37. Interviews, 1985.
38. Kunreuther et al. 1987.
39. Comparative descriptions of compensation schemes can be found in OECD 1980.
40. Morell and Magorian 1982:171.
41. Michelman 1967:1179.
42. Ibid.
43. Morell and Magorian 1982. See also Healy and Rosenburg 1979.
44. Bacow and Wheeler 1991.
45. See Soble and Brennan 1988:1061; and Soble 1977:703.
46. Rabe 1994.

3. Structure of the Bargaining Environment

1. Other information was characterized by missing observations because only census data were available. Missing observations were imputed by computing average annual trend growth during census periods.

2. See Asahi shinbun sha 1960–1982; Jichi shō zaisei kyoku shidō ka 1960–1982; and Sōrifu tōkei kyoku 1960–1982.
3. The *evaluative model* already includes the major results of the *explanatory model*.
4. Calder's (1988) two crisis periods (1958–1963 and 1971–1976) are relevant here.
5. Prefectural electricity self-sufficiency ratios, although positively correlated with settlement times, were not statistically significant. Japan's electricity grid on Honshu is fairly well interconnected, and interprefectural electricity transfers are common. Prefectures are less likely to be willing to incur political fallout from hosting of hazardous facilities if they can get electricity from alternative sources.
6. See Chapter 2.
7. Baumol and Oates 1975:chap. 13.
8. It would be interesting to explore statistically the size of final compensation packages relative to community economic prospects beforehand and to distinguish between high resistance leading to high compensation and high resistance leading to no siting (and no compensation). The unavailability of compensation data for an adequately large number of cases prevented such an investigation.
9. Portney 1991 also finds that in the United States people whose incomes are relatively high are more likely to be willing to accept waste management facilities in their localities, although he offers no careful explanation for this finding.
10. Fossil-fueled plants give off visible air pollution, which may give communities the perception that the risks of those plants are more concentrated than those of nuclear plants. In Japan, the risk of both types of plants in terms of accidents, dangerous emissions, transport, and storage are generally concentrated at the site of combustion. Even if we assume that these risks are perceived by communities as the same for all plants, the visibility of air pollution from fossil-fueled projects may in the mind of communities tip the perceived risk level to the advantage of nuclear power.
11. Communication with Margaret McKean, 1994.
12. There is no meaningful relationship between the strength of Leftist political parties at the local level and public acceptance times, which is not surprising because utilities avoid siting nuclear power plants where there is any significant Leftist party strength.
13. There are at least two reasons for this weighting. The first is that Leftist political parties have often been regarded as counterbalancing the pro-development emphasis of conservative parties. Opposition to nuclear power has been a hallmark of both Japan Communist Party (JCP) and Japan Socialist Party (JSP) policies. The second is that Leftist resistance is also related to opposition to monopoly capitalism, even though the Leftist parties often cast this in terms of resistance because of environmental costs. I am aware of the controversy in the Japan studies literature about the importance of ideology in this context and note that other observers see ideology as not being important to positions on environmental protection. See, for example, Lewis 1980; McKean 1981; and Samuels 1983.
14. See Lesbirel 1991.
15. The dummy variables allow for variations only in the sense of "yes" or "no," whereas other variables allow for much more refined measurement of varia-

tions. In any case, the use of fuel type and package number as proxy measures for the riskiness attached to different projects is debatable. The only real way to test this means of prediction would be to survey local populations.
16. For general analyses that see risk as not being simply determined by the nature of technology, see Starr 1969; Lowrence 1976; Weinberg 1972; Starr 1973; Starr, Rudman, and Whipple 1976; Hohenemser, Kasperson, and Kates 1977; Kunreuther and Linerooth 1982; and Kasperson, Golding, and Tuler 1992. The most comprehensive analysis of project risk can be found in Portney 1991.
17. Based on interviews with personnel from several utilities. The analysis in Chapter 2 also confirms this observation.
18. See Denryoku chūō kenkyū jō 1982 for an excellent quantitative analysis of the impacts of siting on local economies. For American experience, refer to Shields, Cowan, and Bjornstad 1979.
19. Interviews, 1992.

4. Gaining Access to Political Power

1. As discussed in Chapter 3, they included Shimokita, Namie, Noto, Kumano, Hidaka, Suzu, and Hikigawa. Had these projects been accepted, Japan's nuclear capacity would have been more than double what it is today.
2. Interviews, 1984.
3. Data compiled from Shigen enerugii chō, various issues, 1960–1982a.
4. Sato 1978:33–35. See also Nihon genshiryoku sangyō kaigi 1965 for a more detailed analysis of the history of the nuclear power industry in Japan.
5. Interviews, 1984.
6. See Nakamura et al. 1982:10–18.
7. Interviews, 1984.
8. Mie was relying heavily on outside national government finance to develop necessary infrastructure to facilitate economic development. See Jichi shō zaisei kyoku shidō ka, 1964.
9. In 1959, the Ise typhoon devastated extensively ninety public facilities and caused damage worth approximately 218 million yen. In the early 1960s, a tidal wave and a typhoon cost the region approximately ninety million yen in reconstruction. Interviews, 1984.
10. *Chūnichi shinbun* and *Ise shinbun,* various issues, 14–29 November 1963.
11. See Nakamura et al. 1982:20.
12. Interviews, 1984.
13. *Ise shinbun,* various issues, 8–12 March 1964.
14. Interviews 1984.
15. *Ise shinbun* and *Chūbu yomiuri shinbun,* 17 March 1964.
16. Interviews, 1984.
17. Nakamura et al. 1982:23. Nakamura provides a detailed discussion of the resistance by Nanto fishing cooperatives. He was a member of the Kowaura fishing cooperative and was involved in the opposition movement. His analysis focuses on the events in Nanto and argues that resistance by the Central Committee was critical in forcing Chubu Electric to abandon the project. The significance

of this resistance applied to the project at Nanto but, as we shall see, does not work for the project at Kisei.
18. In 1965, there were 1,400 households employed in the Nanto fishing industry. Of these householders, 1,350 were engaged in the production of pearls. Interviews, 1984.
19. *Ise shinbun,* 26 February 1964; and interviews, 1984.
20. Nakamura et al. 1982:25–28.
21. Interviews, 1984.
22. Interviews, 1984.
23. Nakamura et al. 1982:49–51.
24. *Chūnichi shinbun* and *Chūbu yomiuri shinbun,* 14 April 1964.
25. Interviews, 1984.
26. The information in this section was based on extensive interviews in 1983 and 1984.
27. See *Chūnichi shinbun, Ise shinbun,* and *Chūbu shinbun,* various issues, June to August 1964.
28. *Ise shinbun,* 27 July 1964.
29. This section on political factionalism in Kisei was based on extensive interviews in 1983 and 1984.
30. The value of the fishing industry in Kisei was approximately 300 million yen. Interviews, 1983.
31. Interviews, 1984.
32. Prefectural officials felt that the amalgamation of towns with similar industrial structures would enhance administrative efficiency.
33. Nakamura et al. 1982:58–60.
34. *Chūbu yomiuri shinbun,* 18 June 1964.
35. Calculated from data contained in Tsūshō sangyō shō 1965–1971.
36. *Chūnichi shinbun,* 23 July 1965.
37. Interviews conducted in 1984 revealed that Chubu Electric was only prepared to pay compensation for the transfer of property rights in accordance with the Compensation Standards. The figure of 900 million yen was calculated from the formulae contained in the standards (see Chapter 2). Although strategic bargaining considerations obviously would have been involved in arriving at this figure, it does provide a broad indication of the divergence between what the utility was willing to pay and what the Nanto fishing cooperatives were willing to accept to reach settlement over the relinquishment of fishing rights.
38. At that time, Tokyo Electric was siting the Fukushima No. 1 nuclear project and Kansai Electric was developing the Oi nuclear project.
39. Interviews, 1984.
40. Calculated from data contained in Tsūshō sangyō shō, various issues, 1960–1982.
41. Chubu Electric had an agreement with Kansai and Hokuriku Electric, the other two power companies in the central electricity sphere, to purchase electricity in times of expected shortfalls. Interviews, 1984.
42. MITI judged that the utility had not met the necessary criteria to submit the project for national government approval. These criteria included agreement from

local mayors, the prefectural governor, and fishing cooperatives (see Chapter 2). Interviews, 1983.
43. Interviews, 1984.
44. See Nakamura et al. 1982:94–120, 146–150.
45. *Chūnichi shinbun* and *Chūbu yomiuri shinbun,* various issues, 16–18 November 1965.
46. *Ise shinbun,* 21 November 1965; *Chūnichi shinbun,* 22 November 1965; and interviews, 1984.
47. See Nakamura et al. 1982:120–128.
48. After Akira Kurosawa's famous movie of the same name.
49. Interview, 1984.
50. See Nakamura et al. 1982:145–197, for an interesting account of this incident. Also see *Ise shinbun, Chūnichi shinbun,* and *Chūbu yomiuri shinbun,* various issues, September to December 1966.
51. See *Ise shinbun* and *Chūbu yomiuri shinbum,* various issues, January to May 1967.

5. Carving Up Opposition Alliances

1. And, for that matter, faster than many conventional power plants as well. The dispute over Kansai Electric's Oi nuclear plant was the only nuclear dispute resolved more quickly.
2. See *Sankei shinbun,* 5 July 1967.
3. Interviews, 1983, and documents received from Chubu Electric.
4. Interviews, 1983.
5. Interviews, 1984.
6. *Sankei shinbun* and *Shizuoka shinbun,* various issues, August 1967; and interviews, 1984.
7. In 1967, Shizuoka prefecture's electricity self-sufficiency was 85 percent. Nihon enerugii keizai kenkyū jō, 1980.
8. *Asahi shinbun: Shizuoka ban,* 15 August 1967; and interviews, 1983.
9. This designation fell under the Law for the Promotion of Industrial Development. The provisions of this law encouraged industry to locate projects in underdeveloped areas to rectify imbalances in regional economic growth in Japan. The main criteria for designating a region underdeveloped was a Local Financial Index of less than 0.3. Interviews, 1983.
10. Interviews 1983.
11. Jichi shō zaisei kyoku shidō ka 1968; and Hamaoka chō yakuba 1982.
12. Interviews, 1983.
13. Interviews, 1984.
14. *Shizuoka chūnichi shinbun* and *Sankei shinbun,* various issues, April 1967; and interviews, 1984.
15. See *Sankei shinbun,* 4 July 1967; and *Asahi shinbun: Shizuoka ban,* evening edition, 5 July 1967.
16. *Shizuoka chūnichi shinbun,* 16 August 1967; *Asahi shinbun: Enshu ban,* 19

September 1967; and interviews, 1983. See Kōgai taisaku shizuoka ken renraku kaigi (1980:76–91) for a history of the movement opposing the siting of the Hamaoka nuclear power plant. Mori 1982 contains an analysis of the earthquake issue as it relates to the siting of the Hamaoka nuclear power plant.

17. Interviews, 1983.
18. Interviews, 1983.
19. *Shizuoka chūnichi shinbun, Asahi shinbun: Shizuoka ban,* and *Shizuoka shinbun,* various issues, February 1968; and interviews, 1983.
20. Interviews, 1983.
21. See *Shizuoka chūnichi shinbun,* various issues, 4–9 October 1967, and Haraguchi 1973.
22. Interviews, 1983.
23. See Sōrifu tōkei kyoku 1968.
24. Interviews, 1983.
25. *Sankei shinbun* and *Asahi shinbun: Enshu ban,* various issues, September to December 1967, and interviews, 1984.
26. Interviews, 1983.
27. Interviews, 1983.
28. Interviews, 1983.
29. Interviews, 1984.
30. *Shizuoka shinbun,* 7 February 1968; and interviews, 1983.
31. Interviews, 1983.
32. For details of the negotiations, see *Sankei shinbun, Chūnichi shinbun, Shizuoka shinbun,* and *Shizuoka chūnichi shinbun,* various issues, March to July, 1968.
33. Refer to Chapter 2 for a description of the Compensation Standards.
34. *Shizuoka shinbun,* 9 March 1968; *Sankei shinbun,* 10 March 1968; and interviews, 1984.
35. Interviews, 1984.
36. Interviews, 1984.
37. Nihon genshiryoku sangyō kaigi 1965.
38. Interviews, 1984. Also see Lewis 1980.
39. Interviews, 1984. Hamaoka genshiryoku hatsuden shō kensetsu hantai gyōmin kyōgikai 1968 provides an interesting discussion of this development. See also *Shizuoka chūnichi shinbun,* 27 December 1968.
40. Interviews, 1984.
41. See Suzuki 1969 and 1977.
42. Interviews, 1984.
43. Interviews, 1983.
44. *Chūnichi shinbun,* 7–9, 10 July 1969; and interviews, 1984.
45. See Hamaoka genshiryoku hatsuden shō kensetsu hantai gyōmin kyōgikai 1968.
46. Interviews, 1984.
47. Interviews, 1984. On the election, see *Shizuoka shinbun* and *Asahi shinbun: Enshu ban,* various issues, 12–17 August 1969.
48. Interviews, 1984.
49. Shigen enerugii chō, various issues, 1960–1982b.

6. Capitalizing on External Shocks

1. Interviews, 1983.
2. The EPDC also has a role in nuclear research and development and in providing *amakudari* [descent-from-heaven] jobs for retired bureaucrats. See Samuels 1987:239, 245, 160–166.
3. See Ikuta 1984:9–11. From 1960 to 1973, coal production declined from 53 million tonnes to 21 million tonnes, and the number of operating mines declined from 622 to 37. Shigen enerugii chō, various issues, 1960–1982a. Samuels 1987 provides the most detailed history of Japanese coal policy.
4. For a discussion of the origins and structure of subsidy arrangements for protecting the coal industry, see Samuels 1987; and Lesbirel 1994.
5. Interviews, 1983.
6. Interviews, 1983.
7. Interviews, 1985.
8. Interviews, 1983.
9. See Igarashi 1982:7–13. His book, which the EPDC funded, focuses on the company's social responsibility in the siting of the project and stresses the welfare emphasis of the project.
10. Ishikawa 1973.
11. See Kubo 1973.
12. Kyushu Electric's response to the project is based on extensive interviews in 1983 and various issues of *Yomiuri shinbun, Nagasaki shinbun,* and *Nikkan kōgyō shinbun,* 17–18 May 1974. Also refer to Igarashi 1982:68–74.
13. Interviews, 1985.
14. *Nagasaki shinbun,* 22 December 1973. From 1970 to 1973, electricity self-sufficiency had increased from 122 percent to 161 percent. Nihon enerugii keizai kenkyū jō 1980.
15. Information received in interviews, 1984; and Shigen enerugii chō, various issues, 1960–1982b.
16. *Asahi shinbun,* 13 March 1974.
17. *Denki shinbun,* 4 October 1974; and *Nishi nihon shinbun,* 6 October 1974.
18. Interviews, 1983.
19. Interviews, 1983. See also Igarashi 1982:38–57.
20. Interviews, 1983.
21. *Nagasaki shinbun,* 12–14 January 1974; *Nikkan kōgyō shinbun,* 17 January 1974; and interviews, 1983.
22. Omi, ed. 1978:18.
23. *Denki shinbun,* 4 February 1974; and Honma 1981:67–70.
24. Interviews, 1983.
25. See Igarashi 1982:98–104.
26. See Lesbirel 1991.
27. These figures were calculated from information received from interviews in 1983; and Shigen enerugii chō, various issues, 1960–1982b.
28. Interviews, 1983.
29. Interviews, 1983.

30. For various analyses of the dispute, see *Denki shinbun,* 21–25 September 1974; *Nihon keizai shinbun,* 20–22 September 1974; *Nikkan kōgyō shinbun,* 21 September 1974; and *Nishi nihon shinbun,* 20 September 1974.
31. Although Chugoku Electric had a smaller grid, it agreed to take a 40 percent share because of its greater need for electricity. Kyushu Electric agreed to take 40 percent because its grid was the largest in the western sphere. Shikoku Electric had both a smaller grid and less need for electricity and agreed to purchase the remaining 20 percent. Interviews, 1983.
32. See Matsushima richi jimushō 1977:5.
33. Interviews, 1983.
34. *Nihon keizai shinbun,* 13 January 1975.
35. Interviews, 1983.
36. See Igarashi 1982:74–80.
37. *Nihon keizai shinbun,* 13 December 1975. The EPDC's proposal also included the development of a High Temperature Gas Reactor. It requested a budget of 7,750 million yen, of which 820 million yen was for the Matsushima project and 900 million yen was for the transmission network. MOF was willing to provide only 4,030 million yen. Although the company had 1,030 million yen in equity capital, it remained approximately 2,700 million yen below its budget request. This shortfall meant that it could only develop the Matsushima project. But the compromise clearly suited the EPDC, MITI, and the regional electorate in Kyushu. See *Denki shinbun, Nikkan kōgyō shinbun,* and *Nihon keizai shinbun,* various issues, 7–9 January 1976.
38. Igarashi 1982:80–91; and interviews, 1983.
39. See Matsushima richi jimushō 1977:24; and Igarashi 1982:109–111.
40. Interviews, 1983.
41. Interviews, 1983.
42. See Matsushima richi jimushō 1977:22–32.
43. Information contained in an internal EPDC document entitled *Matsushima jiten shūhen chiiki seibi kōfukin haibun* [The Distribution of Compensation for the Development of Public Facilities in Ooseto], 1987.
44. See Dengen kaihatsu kabushiki gaisha 1976.
45. See Matsushima richi jimushō 1977:12.
46. Interviews, 1983.
47. Igarashi 1982:118–124, 145–153.
48. Interviews, 1984.

7. *Dealing with Changing Project Costs*

1. *Denki shinbun,* 31 March 1969.
2. *Hokkaido shinbun,* 28 June 1969.
3. *Shigen enerugii chō,* various issues, 1960–1982a.
4. *Hokkaido shinbun,* 28 June 1969.
5. A power plant that supplies a large share of electricity to a grid reduces reliability because any outage will cause a significant loss of power to the grid. Loss

of power is a major problem for smaller utilities in Japan but does not concern larger utilities, whose grids are relatively large. Interviews, 1983.

6. This section is based on *Nihon keizai shinbun: Hokkaido ban, Nikkan kōgyō shinbun, Hokkaido taimuzu, Asahi shinbun: Hokkaido ban,* and *Hokkaido shinbun,* various issues, 29–30 September 1969; and interviews in 1983.
7. See Tanaka 1981 for an analysis of fishing cooperative and other community resistance in Iwanai town to the project.
8. Interviews, 1984. See also Inuta and Nagatani, eds. 1981:31–64.
9. Interviews, 1983.
10. Interviews, 1983.
11. See *Hokkaido shinbun,* 14 March 1971; and *Hokkaido taimuzu,* 14 March 1971.
12. Hokkaido shōkō kankyō bu 1983:161–162; and interviews, 1983.
13. Asahi shinbun sha, various issues, 1960–1982; and interviews, 1983.
14. Jichi shō zaisei kyoku shidō ka, various issues, 1960–1982; and interviews, 1983.
15. See *Asahi shinbun: Hokkaido ban, Mainichi shinbun: Hokkaido ban, Hokkaido taimuzu,* and *Hokkaido shinbun,* various issues, 21 February to 9 March 1973.
16. Interviews, 1983.
17. See *Hokkaido shinbun,* 16 July 1973.
18. Interviews, 1983.
19. Shigen enerugii chō, various issues, 1960–1982a; and interviews, 1983.
20. This section on scheduling the Kyowa-Tomari and Date projects is based on extensive interviews, 1983.
21. See Hokkaido shōkō kankyō bu (1983:109–123) for a more detailed statement of Hokkaido's energy policy.
22. Interviews, 1983.
23. Interviews, 1983.
24. This section on Hokkaido Electric's strategies for isolating the Iwanai fishing cooperative is based on interviews, 1983.
25. Interviews, 1983.
26. The electricity spheres of all the other utilities in Japan, except Okinawa Electric, comprise more than one prefecture.
27. See an interesting supplement in *Nihon keizai shinbun,* 4 December 1978 for details.
28. *Hokkaido shinbun,* 9 October 1976.
29. See *Nihon keizai shinbun: Hokkaido ban, Hokkaido taimuzu,* and *Asahi shinbun: Hokkaido ban,* various issues, 20–21 April 1972; and *Asahi shinbun, Hokkaido taimuzu, Mainichi shinbun: Hokkaido ban, Yomiuri shinbun: Hokkaido ban,* and *Hokkaido shinbun: Shiribeshi ban,* various issues, 19–21 August 1972.
30. Interviews, 1983.
31. Interviews, 1983.
32. See *Hokkaido taimuzu* and *Hokkaido shinbun,* various issues, 9–13 February 1976.
33. Interviews, 1983.
34. For details, see *Hokkaido taimuzu, Asahi shinbun: Hokkaido ban, Mainichi shinbun: Hokkaido ban, Yomiuri shinbun: Hokkaido ban,* and *Hokkaido shinbun,* various issues, 12–17 March 1977.

35. See *Mainichi shinbun: Hokkaido ban, Yomiuri shinbun: Hokkaido ban,* and *Hokkaido taimuzu,* various issues, 4–6 March 1978.
36. Interviews, 1983.
37. Interviews, 1983.
38. Interviews, 1983.
39. Shigen enerugii chō, various issues, 1960–1982b; and *Hokkaido taimuzu,* 12 October 1978.
40. See *Hokkaido shinbun* and *Hokkaido taimuzu,* various issues, 30 September to 1 October 1978.
41. Interviews, 1983.
42. See Shigen enerugii chō, various issues, 1968–1982b.
43. See *Hokkaido shinbun,* 2 December 1978; and *Mainichi shinbun: Hokkaido ban,* 1 February 1979.
44. Interviews, 1983.
45. For a discussion of the events leading up to the negotiations and the actual negotiations, see *Hokkaido shinbun, Hokkaido taimuzu, Yomiuri shinbun: Hokkaido ban, Mainichi shinbun: Hokkaido ban,* and *Nihon keizai shinbun: Hokkaido ban,* various issues, May to June 1981.
46. *Asahi shinbun: Hokkaido ban,* 27 June 1981; *Hokkaido shinbun, Asahi shinbun: Hokkaido ban,* and *Yomiuri shinbun: Hokkaido ban,* 1 March 1982.
47. Interviews, 1983.

8. NIMBY Politics in Japan

1. Singer 1980.
2. Samuels 1987.
3. Mouer and Sugimoto 1986; and Van Wolferen 1989.
4. Samuels 1987:20–21.
5. Fisher and Ury 1983.
6. O'Hare 1977; and O'Hare, Sanderson, and Bacow 1983.
7. Krauss, Rohlen, and Steinhoff, eds. 1984. Pharr 1990; and Calder 1988b would similarly agree.
8. Pharr 1990:12.
9. Calder 1988b.
10. Kunreuther and Easterling 1991.
11. Portney 1991.
12. See IEA 1996 for historical data on electricity demand growth rates.
13. Baumol and Oates 1975:chap. 13.
14. Aqua 1980.
15. Morell and Magorian 1982:108. See also Singer 1980.
16. Samuels 1983. See also Aqua 1980; Lewis 1980; and McKean 1981.
17. Portney (1991:81–85) finds that Americans who are Republicans are more likely than those who are Democrats to oppose waste treatment facilities in their vicinity.
18. Casper and Wellstone 1981 show how conservative farmers in the United States can influence siting processes over the development of transmission lines.

19. Franzen 1976.
20. For excellent discussions of coalition behavior, see Kelley 1968; and Groennings, Kelley, and Leiserson, eds. 1970.
21. See McKean (1976:234) for a discussion of temporary alliances.
22. See Pharr 1990; and Calder 1988b.
23. See Silberman and Harootunian, eds. 1966.
24. Krauss and Simcock 1980; and McKean 1981.
25. See Maki genpatsu hantai chōmin kaigi 1997 for a history of the Maki siting controversy.
26. *Nikkei Weekly,* 12 August 1997.
27. See Evans and Humphrey 1997; and Lesbirel 1997.
28. Samuels 1983. Several other studies find that leadership matters in resolving conflict in Japan more generally. See Silberman and Harootunian, eds. 1966; McKean 1976; Lewis 1980; McKean 1981; MacDougall, ed. 1982; and Pempel 1982.
29. Cordes and Weisbrod 1985:179.
30. See Lipsky 1971; Von Meter and Von Horn 1975; Sabatier and Mazmanian 1980; Browning, Marshall, and Tabb 1981; and Hjorn and Hull 1982.
31. Portney 1991:9.
32. See, for example, Campbell 1977a; McKean 1981; and McCraw, ed. 1986.
33. For some excellent studies that develop mixed comparisons, see Pempel 1979 and 1982; Johnson 1982; Dore 1986; and Samuels 1987.
34. See O'Hare 1977; and O'Hare, Sanderson, and Bacow 1983.

References

Allinson, Gary, D. and Yasunori Sone, eds. 1993. *Political Dynamics in Contemporary Japan.* Ithaca: Cornell University Press.
Andrews, Richard N. L, and Terrence K. Pierson. 1985. "Local Control or State Override: Experiences and Lessons to Date." *Policy Studies Journal* 14 (Fall): 377–386.
Aoki, Masahiko. 1988. *Information, Incentives, and Bargaining in the Japanese Economy.* Cambridge: Cambridge University Press.
Apter, David E., and Nagayo Sawa. 1984. *Against the State: Politics and Social Protest in Japan.* Cambridge: Harvard University Press.
Aqua, Ronald. 1980. "Political Choice and Policy Change in Medium-Sized Japanese Cities, 1962–1974." In *Political Opposition and Local Politics in Japan,* ed. Kurt Steiner, Ellis S. Krauss, and Scott C. Flanagan. Princeton: Princeton University Press.
Asahi shinbun: enshu ban [Asahi newspaper: Enshu edition].
Asahi shinbun: hokkaido ban [Asahi newspaper: Hokkaido edition].
Asahi shinbun: shizuoka ban [Asahi newspaper: Shizuoka edition].
Asahi shinbun sha. Various issues, 1960–1982. *Asahi nenkan* [Asahi yearbook]. Tokyo: Asahi shinbun sha.
Austin, Murray, Tony E. Smith, and Julian Wolpert. 1970. "The Implementation of Controversial Facility-Complex Programs." *Geographical Analysis* 2: 315–329.
Bacow, Lawrence S. 1980. "Creating Markets for Development Externalities." MIT Energy Laboratory Working Paper, MIT-EL 80-030WP. Cambridge: MIT Energy Laboratory.
Bacow, Lawrence S., and Debra R. Sanderson. 1980. "Facility Siting and Compensation: A Handbook for Communities and Developers." MIT Energy Laboratory Working Paper, MIT-EL 80-037WD. Cambridge: MIT Energy Laboratory.
Bacow, Lawrence S., and Michael Wheeler. 1991. "Binding Parties to Agreements in Environmental Disputes." *Villanova Environmental Law Journal* 2, 1: 99–109.

Baram, Michael. 1976. *Environmental Law and the Siting of Facilities: Issues in Land Use and Coastal Zone Management.* Cambridge: Ballinger.

Barrett, Susan, and Michael Hill. 1984. "Policy, Bargaining, and Structure in Implementation Theory: Towards an Integrated Perspective." *Policy and Politics* 12, 3: 219–240.

Baumol, William J., and Wallace E. Oates. 1975. *The Theory of Environmental Policy: Externalities, Public Outlays, and Environmental Policy.* Englewood Cliffs, N.J.: Prentice Hall.

Bradford, David, and Harold Feiveson. 1976. "Benefits and Costs, Winners and Losers." In *Boundaries of Analysis,* ed. Harold Feiveson, Frank Sinden, and Robert Socolow. Cambridge: Ballinger.

Broadbent, Jeffery. 1986. "The Ties that Bind: Local Social Fabric and the Mobilization of Environmental Movements in Japan." *International Journal of Mass Emergencies and Disaster* 4, 2: 227–253.

Browning, Rufus, Dale Marshall, and David Tabb. 1981. "Implementation and Political Change: Sources of Local Variation in Federal Social Programs." In *Effective Policy Implementation,* ed. Paul Sabatier and Daniel Mazmanian. Lexington: D. C. Heath.

Calder, Kent E. 1988a. "Japanese Foreign Economic Policy Formulation: Explaining the Reactive State." *World Politics* 40: 517–541.

——. 1988b. *Crisis and Compensation: Public Policy and Political Stability in Japan 1949–1986.* Princeton: Princeton University Press.

Campbell, John Creighton. 1977a. *Contemporary Japanese Budget Politics.* Berkeley: University of California Press.

——. 1977b. "Compensation for Repatriates: A Case Study of Interest-Groups Politics and Party-Government Negotiations in Japan." In *Policymaking in Contemporary Japan,* ed. T. J. Pempel. Ithaca: Cornell University Press.

Casper, Barry, and Paul David Wellstone. 1981. *Powerline: The First Battle of America's Energy War.* Amherst: University of Massachusetts Press.

Chūbu yomiuri shinbun [Central yomiuri newspaper].

Chūnichi shimbun [Chunichi newspaper].

Condron, Margaret M., and Dixie L. Sipher. 1983. *Hazardous Waste Facility Siting: A National Survey.* Albany, NY: Legislative Commission on Toxic Substances and Hazardous Waste.

Cordes, Joseph J., and Burton A. Weisbrod. 1985. "When Government Programs Create Inequities: A Guide to Compensation Policies." *Journal of Policy Analysis and Management* 4, 2: 178–195.

Costonis, John J. 1975. "'Fair' Compensation and the Accommodation of Power: Antidotes for the Taking Impasse in Land Use Controversies." *Columbia Law Review* 75, 6: 1021–1082.

Dengen kaihatsu kabushiki gaisha. 1976. *Matsushima denpatsu kensetsu ni taisuru matsushima chiiku jumin no jōken yōbō* [Community demands and conditions regarding the Matsushima coal-fired plant]. Tokyo.

Denki shinbun [Electricity newspaper].

Denryoku chūō kenkyū jō. 1982. *Hatsuden sho richi ni tomonau shakai kankyō henka* [Social environmental impacts accompanying power plant siting in Japan]. Tokyo.

Donnelly, Michael W. 1977. "Setting the Price of Rice: A Study of Political Decisionmaking." In *Policymaking in Contemporary Japan,* ed. T. J. Pempel. Ithaca: Cornell University Press.

Dore, Ronald P. 1986. *Flexible Rigidities: Industrial Policy and Structural Adjustment in the Japanese Economy.* Stanford: Stanford University Press.

Ducsik, Dennis W., ed. 1986. *Public Involvement in Energy Facility Planning: The Electric Utility Experience.* Boulder: Westview Press.

Eisenstadt, S. N., and Eyal Ben-Ari. 1990. *Japanese Models of Conflict Resolution.* London: Kegan Paul International.

Environmental Planning Agency. 1979. *Siting of Hazardous Waste Management Facilities and Public Opposition.* Washington, DC.

Ervin, David, and James Fitch. 1979. "Evaluating Alternative Compensation and Recapture Techniques for Expanded Public Control of Land Use." *Natural Resources Journal* 19 (January): 21–41.

Evans, Peter, and Bruce Humphrey. 1997. "Japan's Deregulated Power Market: Taking Shape." A CERA Global Power Forum Report. Cambridge: Cambridge Energy Research Associates.

Fisher, Roger, and William Ury. 1983. *Getting to Yes: Negotiating Agreement without Giving In.* New York: Penguin Books.

Franzen, F. S. 1976. "Socio-economic Issues for Nuclear Plants: The German Situation." *Trans American Nuclear Society* 24: 95–104.

Friedman, David. 1983. "Beyond the Age of Ford: The Strategic Basis of the Japanese Success in Automobiles." In *American Industry in International Competition: Government Policies and Corporate Strategies,* ed. John Zysman and Laura Tyson. Ithaca: Cornell University Press.

Fukui, Haruhiro. 1977. "Studies in Policymaking: A Review of the Literature." In *Policymaking in Contemporary Japan,* ed. T. J. Pempel. Ithaca: Cornell University Press.

Gamson, William A. 1975. *The Strategy of Social Protest.* Homewood, Ill. Dorsey Press.

George, Aurelia. 1982. "The Comparative Study of Interest Groups in Japan: An Institutional Framework." *Pacific Economic Papers,* no. 95. Canberra: Australia-Japan Research Centre.

Gladwin, Thomas N. 1980. "Patterns of Environmental Conflict over Industrial Facilities in the United States, 1970–78." *Natural Resources Journal* 20 (April): 243–274.

Granger, John A., and Kenneth. R. Wise. 1980. "A Critique of One-Stop Siting: Streamlining Review without Compromising Effectiveness." *Environmental Law* 10 (Summer): 457–482.

Groennings, Sven, E. W. Kelley, and Michael Leiserson, eds. 1970. *The Study of Coalition Behavior: Theoretical Perspectives and Cases from Four Continents.* New York: Holt, Rinehart and Winston.

Gresser, Julian, Koichiro Fujikura, and Akio Morishima. 1981. *Environmental Law in Japan.* Cambridge: MIT Press.

Hadden, Susan, and Jared Hazelton. 1980. "Public Policies Toward Risk." *Policy Studies Journal.* 9,1: 109–117.

Haley, John O. 1987. "Governance by Negotiation: A Reappraisal of Bureaucratic Power in Japan." *Journal of Japanese Studies* 13 (Summer): 343–357.

Hamaoka chō yakuba. 1982. *Hamaoka: Shizuoka ken hamaoka chō chōsei yōran* [Hamaoka town: A survey]. Hamaoka.
Hamaoka genshiryoku hatsuden sho kensetsu hantai gyōmin kyōgikai. 1968. *Hamaoka genpatsu ni taisuru gyōmin no toitsu kenkai* [A united fishermen's view of the Hamaoka nuclear power plant]. Sagara.
Hamilton, Michael S. 1979. "Power Plant Siting: A Literature Review." *Natural Resources Journal* 19, 1: 75–95.
Hansen, Susan B. 1984. "On the Making of Unpopular Decisions: A Typology and Some Evidence." *Policy Studies Journal*. 13, 1: 23–43.
Haraguchi Inoue. 1973. *Gyōgyō to genpatsu* [The fishing industry and nuclear power]. Mie: Kumano genpatsu chōsa kenkyū kai.
Healy, Robert, and John Rosenburg. 1979. *Land Use and the States*. 2nd ed. Baltimore: Johns Hopkins University Press.
Hettich, W. 1969. "Distribution in Cost-Benefit Analysis: A Review of Theoretical Issues." *Public Finance Quarterly* 4 (December): 123–150.
Hjorn, Benny, and Chris Hull. 1982. "Implementation Research as Empirical Constitutionalism." *European Journal of Political Research* 10 (June): 105–116.
Hohenemser, Christoph, Roger Kasperson, and Robert Kates. 1977. "The Distrust of Nuclear Power." *Science* 196: 25–34.
Hokkaido shinbun [Hokkaido newspaper].
Hokkaido shinbun: shiribeshi ban [Hokkaido newspaper: Shiribeshi edition].
Hokkaido shōkō kankyō bu. 1983. *Hokkaido enerugii gaiyō* [Energy in Hokkaido]. Hokkaido.
Hokkaido taimuzu [Hokkaido Times].
Honma, U. 1981. "Machi ni waka tachi ga kaette kita" [The youth have returned]. *Enerugii fuoramu* [Energy forum] 47 (February): 31–36.
Huddle, Norie, and Michael Reich. 1975. *Island of Dreams: Environmental Crisis in Japan*. New York: Autumn Press.
IEA. 1996. *Energy Statistics of OECD Countries*. Paris: OECD.
Igarashi, T. 1982. *Kojima ni dekita hatsuden sho: Denpatsu matsushima sekitan karyoku* [Constructing a power plant on a small island: The matsushima coal-fired plant]. Tokyo: Nikkei jigyō.
Ikuta, T. 1984. "Japan Energy Policies." *Enerugii keizai* [Journal of Energy Economics] 9 (October): 1–18.
Inaba, Y. 1977. *Paburikku akuseputansu* [Public acceptance]. Tokyo: Nihon denki kyōkai.
Inoguchi, T., and T. Iwai. 1987. *Zoku giin no kenkyū: Jimintō seiken o gyūjiru shuyakutachi* [Research on Japanese tribal diet members: The leading players who command the LDP]. Tokyo: Nihon keizai kenkyū jō.
Inuta, T., and K. Nagatani, eds. 1981. *Chiiki funsō no kenkyū: Jichi tai no yakuwari to gōi keisei no jōken* [Research on regional conflict: The role of local administration and conditions for public acceptance]. Tokyo: Gakuyō shōbō.
Ise shinbun [Ise newspaper].
Ishikawa, S. 1973. *Sekitan senkō karyoku hatsuden sho no kensetsu ni taisuru yōboho* [Community demands regarding the development of the Matsushima fossil-fueled plant]. Nagasaki: Nagasaki ken tangyō shichōson rengō kai.
Jichi shō zaisei kyoku shidō ka. Various issues, 1960–1982. *Shichōson betsu kessan jōkyō shirabe* [An examination of local financial accounts]. Tokyo.

Johnson, Chalmers. 1982. *MITI and the Japanese Miracle: The Growth of Industrial Policy, 1925–1975.* Stanford: Stanford University Press.

Jopling, David G. 1974. "Plant Site Evaluation Using Numerical Ratings." *Power Engineering* 4 (March): 56–59.

Jopling, David G., Stephen J. Gage, and Milton E. F. Schoeman. 1973. "Forecasting Public Resistance to Technology: The Example of Nuclear Power Reactor Siting." In *A Guide to Practical Technology Forecasting,* ed. James R. Bright and Milton E. F. Shoeman. Englewood Cliffs, N. J.: Prentice Hall.

Kabashima, Ikuo, and Jeffrey Broadbent. 1986. "Referent Pluralism: Mass Media and Politics in Japan." *Journal of Japanese Studies* 12, 2: 329–361.

Kasperson, R. E., D. Golding, and S. Tuler. 1992. "Siting Hazardous Facilities and Communicating Risks under Conditions of High Social Distrust." *Journal of Social Issues* 48: 161–167.

Keeney, Ralph. 1980. *Siting Energy Facilities.* New York: Academic Press.

Kelley, E. W. 1968. "Techniques for Studying Coalition Formation." *Midwest Journal of Political Science* 12, 1: 62–84.

Knuth, Donald F, and John E. McEwen. 1977. "Trends in the Licensing of Nuclear Power Plants." *Nuclear Safety* 18 (September/October): 581–588.

Kobayashi, T. 1983. *Kōkyō yōchi no shutoku ni tomonau sonshitsu hoshō kijun yōko no kaisetsu* [An analysis of the compensation standards accompanying the acquisition of public lands in Japan]. Tokyo: Kindai tosho kabushiki gaisha.

Kōgai taisaku shizuoka ken renraku kaigi. 1980. *Kōgai to shizuoka kenmin* [Pollution and the Shizuoka community]. no. 6. Shizuoka.

Krauss, Ellis S, Thomas P. Rohlen, and Patricia G. Steinhoff, eds. 1984. *Conflict in Japan.* Honolulu: University of Hawaii Press.

Krauss, Ellis S., and Bradford L. Simcock. 1980. "Citizens' Movements: The Growth and Impact of Environmental Protest in Japan." In *Political Opposition and Local Politics in Japan,* ed. Kurt Steiner, Ellis S. Krauss, and Scott C. Flanagan. Princeton: Princeton University Press.

Kretzmer, D. 1979. "Legal Problems of Binding Communities to Compensation Agreements for Adverse Effects of Energy Facilities." Cambridge: MIT Laboratory of Architecture and Planning.

Kubo, S. 1973. *Sekitan senkō karyoku hatsuden sho kensetsu ni kansuru yōbōsho* [Demands regarding the construction of the Matsushima coal-fired project]. Nagasaki: Dengen kaihatsu kabushiki gaisha.

Kunreuther, Howard, and Douglas Easterling. 1991. "Are Risk-Benefit Tradeoffs Possible in Siting Hazardous Facilities?" *American Economic Review* 33: 252–256.

Kunreuther, Howard, et al. 1987. "A Compensation Mechanism for Siting Noxious Facilities: Theory and Experimental Design." *Journal of Environmental Economics and Management* 14, 4: 371–383.

Kunreuther, Howard, and Joanne Linerooth. 1982. *Risk Analysis and Decision Processes: The Siting of Liquified Energy Facilities in Four Countries.* Berlin: Springer-Verlag.

Kunreuther, Howard, Paul Slovic, and Donald MacGregor. 1996. "Risk Perception and Trust: Challenges for Facility Siting." *Risk: Health, Safety, and Environment* 7, 2: 109–118.

Lowrance, William W. 1976. *Of Acceptable Risk: Science and the Determination of Safety.* Los Altos: William Kaufman.
Lesbirel, S. Hayden. 1988. "The Political Economy of Substitution Policy: Japan's Response to Lower Oil Prices." *Pacific Affairs* 61, 2: 285–302.
———. 1991. "Structural Adjustment in Japan: Terminating 'Old King Coal.' " *Asian Survey* 31, 11: 1079–1094.
———. 1994. "Risk Sharing Mechanisms and Policy Implementation: Structural Adjustment in the Japanese Coal Industry in Comparative Perspective." *Asian Journal of Political Science* 2, 2: 89–111.
———. 1997. "Wheeling and Dealing: Reforming Electricity Markets in Japan." *MIT Japan Papers* MITJP#97-01, Cambridge: Center for International Studies, MIT.
Lester, Richard K. 1983. "Nuclear Power Plant Leadtimes." In *World Nuclear Energy: Toward a Bargain of Confidence,* ed. Ian Smart. Baltimore: Johns Hopkins University Press.
Lewis, Jack G. 1980. "Civic Protest in Mishima: Citizens' Movements and the Politics of the Environment in Contemporary Japan." In *Political Opposition and Local Politics in Japan,* ed. Kurt Steiner, Ellis S. Krauss, and Scott C. Flanagan. Princeton: Princeton University Press.
Lipsky, Michael. 1968. "Protest as a Political Resource." *American Political Science Review* 67 (December): 1144–1158.
———. 1971. "Street Level Bureaucracy and the Analysis of Urban Reform." *Urban Affairs Quarterly* 6: 391–401.
MacDougall, Terry Edward, ed. 1982. "Political Leadership in Contemporary Japan." *Michigan Papers in Japanese Studies* no. 1, Michigan: Center for Japanese Studies, University of Michigan.
Mainichi shinbun: hokkaido ban [Mainichi newspaper: Hokkaido edition].
Maki genpatsu hantai chōmin kaigi. 1997. *Maki genpatsu o meguru jōsei* [The Maki nuclear plant: Current developments]. Niigata.
Matsushima richi jimushō. 1977. *Matsushima karyoku hatsuden sho kensetsu keikaku: Keii to genjō* [The construction of the Matsushima power plant: Details and the present situation]. Tokyo.
McCraw, Thomas K., ed. 1986. *America versus Japan.* Cambridge: Harvard Business School Press.
McKean, Margaret A. 1976. "Citizens' Movements in Urban and Rural Japan." In *Social Change and Community Politics in Urban Japan,* ed. James W. White and Frank Munger. Comparative Urban Studies Monograph No. 4. Chapel Hill: Institute for Research in Social Science, University of North Carolina at Chapel Hill.
———. 1981. *Environmental Protest and Citizen Politics in Japan.* Berkeley and Los Angeles: University of California Press.
———. 1993. "State Strength and the Public Interest in Japan" In *Political Dynamics in Contemporary Japan,* ed. Gary Allison and Yasunori Sone. Ithaca: Cornell University Press.
McMahon, Robert, Cindy Ernst, Ray Miyares, and Curtis Haymore. 1982. *Using Compensation and Incentives When Siting Hazardous Waste Management Facilities—A Handbook.* Washington, D.C.: U.S. Environmental Planning Agency.
Michelman, Frank I. 1967. "Property, Utility and Fairness: Comments on the Ethical Foundations of 'Just Compensation' Law." *Harvard Law Review* 80, 6: 1165–1258.

Mishan, Edward J. 1972. *Cost-Benefit Analysis: An Informal Introduction.* London: Allen and Unwin.

Mizumoto, H. 1980. *Tochi mondai to shoyūken* [Land problems and land rights in Japan]. Tokyo: Yudanraku sensho.

Morell, David, and Christopher Magorian. 1982. *Siting Hazardous Waste Facilities: Local Opposition and the Myth of Pre-emption.* Cambridge: Ballinger.

Mori, S. 1982. *Genpatsu no machi kara: Tokai daijishintai jō no hamaoka genpatsu* [From a nuclear town: The Hamaoka nuclear power plant located on the Tokai fault]. Tokyo: Hatatashoten.

Mouer, Ross E, and Yoshio Sugimoto. 1986. *Images of Japanese Society: A Study of the Social Construction of Reality.* London: Kegan Paul International.

Mumphrey, Anthony J., and Julian Wolpert. 1973. "Equity Considerations and Concessions in the Siting of Public Facilities." *Economic Geography* 48: 109–121.

Muramatsu, Michio. 1986. "Center-Local Political Relations in Japan: A Lateral Competition Model." *Journal of Japanese Studies* 12 (Summer): 303–327.

Muramatsu, Michio, and Ellis Krauss. 1987. "The Conservative Policy Line and the Development of Patterned Pluralism." In *The Political Economy of Japan, Vol. 1: The Domestic Transformation,* ed. Kozo Yamamura and Yasukichi Yasuba. Stanford: Stanford University Press.

Murota, Yasuhiro. 1984. *Enerugii* [Energy]. Tokyo: Kyōikusha.

Murray, William, and Carl Seneker. 1980. "Implementation of an Industrial Siting Plan." *Hastings Law Journal* 3 (May): 1073–1089.

Nagasaki shinbun [Nagasaki newspaper].

Najita, Tetsuo, and J. Victor Koschmann, eds. 1982. *Conflict in Modern Japanese History: The Neglected Tradition.* Princeton: Princeton University Press.

Nakamura, K., et al. 1982. *Genpatsu: Kumano gyōmin kaisenki.* [Nuclear power: The sea battle at Kumano]. Tokyo: Gijutsu to ningen.

Nakatani, I. 1984. "The Economic Role of Financial Corporate Grouping." In *The Economic Analysis of the Japanese Firm,* ed. Masahiko Aoki. Amsterdam: North Holland.

Nash, C., D. Pearce, and J. Stanley. 1975. "An Evaluation of Cost-Benefit Analysis Criteria." *Scottish Journal of Political Economy* 22: 119–134.

Nelkin, Dorothy. 1977. *Technological Decisions and Democracy: European Experiments in Public Participation.* Beverly Hills: Sage Publications.

Nelkin, Dorothy, and Michael Pollack. 1981. *The Atom Besieged: Extra-Parliamentary Dissent in France and West Germany.* Cambridge: MIT Press.

Nihon denki kyōkai. 1981. *Anata no shiritai koto* [All you need to know about the electric power industry in Japan]. Tokyo.

Nihon enerugii keizai kenkyū jō. 1980. *Chiiki betsu enerugii juyō tokusei no bunseki* [Characteristics of energy supply and demand in Japan by prefecture]. Tokyo.

Nihon genshiryoku sangyō kaigi. 1965. *Genshiryoku kaihatsu ju nenshi* [A ten year history of nuclear power in Japan]. Vols 1 and 2. Tokyo.

———. 1989. *Genshiryoku hatsuden sho ranhyō* [Data on world nuclear power plants]. Tokyo.

Nihon keizai shinbun [Japan economic journal].

Nihon keizai shinbun: hokkaido ban [Japan economic journal: Hokkaido edition].

Nikkan kōgyō shinbun [Japan industrial newspaper].

Nikkei Weekly.
Nishi nihon shinbun [Western Japan newspaper].
OECD. 1975. *Siting of Nuclear Facilities: Proceedings of a Symposium.* (9–12 December). Vienna: IAEA.
———. 1980. *Siting Procedures for Major Energy Facilities: Some National Cases.* Paris.
O'Hare, Michael. 1977. "Not on My Block You Don't. . . . Facility Siting and the Strategic Importance of Compensation." *Public Policy* 24, 4: 407–458.
O'Hare, Michael, Debra Sanderson, and Lawrence Bacow. 1983. *Facility Siting and Public Opposition.* New York: Van Nostrand-Reinhold.
Okimoto, Daniel I. 1989. *Between MITI and the Market: Japanese Industrial Policy for High Technology.* Stanford: Stanford University Press.
Olson, Mancur. 1965. *The Logic of Collective Action.* Cambridge: Harvard University Press.
Omi, K., ed. 1978. "Matsushima sekitan karyoku hatsuden sho richi o kaerimiru" [Reflecting on the development of the Matsushima coal-fired plant]. *Denpatsu* (January): 45–48.
Pearce, David W. 1971. *Cost-Benefit Analysis.* London: MacMillan.
Pempel, T. J. 1982. *Policy and Politics in Japan: Creative Conservatism.* Philadelphia: Temple University Press.
———. 1987. "The Unbundling of 'Japan, Inc.' The Changing Dynamics of Japanese Policy Formulation." *Journal of Japanese Studies* 13 (Summer): 271–306.
———. ed. 1977. *Policymaking in Contemporary Japan.* Ithaca: Cornell University Press.
Pempel, T. J., and K. Tsunekawa. 1979. "Corporatism without Labor? The Japanese Anomaly." In *Trends toward Corporatist Intermediation,* ed. P. C. Schmitter and G. Lembruch. Beverly Hills: Sage Publications.
Pharr, Susan. 1990. *Losing Face: Status Politics in Japan.* Berkeley: University of California Press.
Popper, F. 1983. "LP/HC and LULUs: The Political Uses of Risk Analysis in Land Use Planning." *Risk Analysis* 3, 4: 255–263.
Portney, Kent E. 1991. *Siting Hazardous Waste Facilities: The Nimby Syndrome.* New York: Auburn House.
Pressman, Jeffrey L., and Aaron Wildavsky. 1974. *Implementation: How Great Expectations in Washington Are Dashed in Oakland.* Berkeley: University of California Press.
Prestowitz, Clyde. 1988. *Trading Places.* New York: Basic Books.
Pringle, Peter, and James Spingelman. 1981. *The Nuclear Barons.* New York: Holt, Rinehart and Winston.
Quirk, James, and Katsuaki Terasawa. 1981. "Nuclear Regulation: An Historical Perspective." *Natural Resources Journal* 21 (October): 833–855.
Rabe, Barry G. 1994. *Beyond NIMBY: Hazardous Waste Siting in Canada and the United States.* Washington, D.C.: Brookings Institution.
Rad, Parviz F. 1979. "Delays in Construction of Nuclear Power Plants." *Journal of Energy Division* 105, 1: 33–46.
Radlauer, Marcy A., David S. Bauman, and Stephen W. Chapel. 1985. "Nuclear Construction Leadtimes: Analysis of Past Trends and Outlook for the Future." *Energy Journal* 6 (January): 45–62.

Raiffa, H. 1982. *The Art and Science of Negotiations.* Cambridge: Harvard University Press.
Ramseyer, Mark, and Frances Rosenbluth. 1993. *Japan's Political Marketplace.* Princeton: Princeton University Press.
Reed, Krista S., and C. Edwin Young. 1983. "Impact of Regulatory Delays on the Cost of Wastewater Treatment Plants." *Land Economics* 59, 1: 35–42.
Reed, Steven R. 1982. "Is Japanese Government Really Centralized?" *Journal of Japanese Studies* 8 (Summer): 133–164.
———. 1986. *Japanese Prefectures and Policymaking.* Pittsburgh: Pittsburgh University Press.
Reich, Michael. 1991. *Toxic Politics: Responding to Chemical Disasters.* Ithaca: Cornell University Press.
Roberts, Marc J. 1981. *The Choices of Power: Utilities Face the Environmental Challenge.* Cambridge: Harvard University Press.
Rosenbluth, Frances M. 1989. *Financial Politics in Contemporary Japan.* Ithaca: Cornell University Press.
Sabatier, Paul, and Daniel Mazmanian. 1979. "The Conditions of Effective Implementation." *Policy Analysis* 5 (Fall): 481–504.
———. 1980. "A Framework for Analysis." *Policy Studies Journal* 8: 538–560.
Samuels, Richard J. 1983. *The Politics of Regional Policy in Japan: Localities Incorporated?* Princeton: Princeton University Press.
———. 1987. *The Business of the Japanese State: Energy Markets in Comparative and Historical Perspective.* Ithaca: Cornell University Press.
Sankei shinbun [Sankei Newspaper].
Sato, H. 1978. *The Politics of Technology Importation in Japan: The Case of Atomic Power Reactors.* Paper delivered to SSRC Workshop at Kona, Hawaii, 7–11 February, 1978.
Sato, T. 1978. *Nihon gyōgyō no horitsu mondai* [Legal problems of the Japanese fishing industry]. Tokyo: Keiso shōbō.
Schattschneider, Edwin E. 1960. *The Semi-Sovereign People.* New York: Holt, Rinehart and Winston.
Schelling, Thomas C. 1960. *The Strategy of Conflict.* New York: Oxford University Press.
Shakai keizai kokumin kaigi. 1981. *Enerugii kanren shisetsu no richi taisaki* [Siting policies for energy-related facilities]. Tokyo: Shakai keizai kokumin kaigi.
Sheard, Paul. 1991. "The Role of Firm Organization in the Adjustment of Declining Industry in Japan: The Case of Aluminum." *Journal of Japanese and International Economies* 5, 1: 19–40.
Shields, Mark A., J. Tadlock Cowan, and David J. Bjornstad. 1979. *Socioeconomic Impacts of Nuclear Power Plants: A Paired Comparison of Operating Facilities.* NUREG/CR-0916. Oak Ridge: Oak Ridge National Laboratory.
Shigen enerugii chō. Various issues, 1960–1982a. *Dengen kaihatsu no gaiyō: Sono keikaku to kisō shiryō* [An outline of power plant development in Japan: Plans and basic data]. Tokyo.
———. Various issues, 1960–1982b. *Sōgō enerugii tōkei* [Energy statistics]. Tokyo.
———. 1985. *Enerugii roppō* [The six energy laws]. Tokyo: Toyō hoki shuppan.
———. 1987. *Dengen sanpō no gaiyō* [An outline of the three laws]. Tokyo.

Shin denki jigyō kōza. 1980a. *Denki jigyō hattatsu shi* [A history of the development of the electric power industry]. Tokyo: Denryoku shinpō sha.

——. 1980b. *Denryoku keitō keikaku to unyō* [Planning and operating electricity systems]. Tokyo: Denryoku shinpō sha.

Shizuoka chūnichi shinbun [Shizuoka central newspaper].

Shizuoka shinbun [Shizuoka newspaper].

Silberman, Bernard S., and H. D. Harootunian, eds. 1966. *Modern Japanese Leadership: Transition and Change*. Tucson: University of Arizona Press.

Singer, Grace. 1980. "People and Petrochemicals: Siting Controversies on the Waterfront." In *Refining the Waterfront: Alternative Energy Policies for Urban Coastal Areas,* ed. David Morell and Grace Singer. Cambridge: Oelgeschlager, Gunn and Hain.

Soble, Stephen M. 1977. "A Proposal for Administrative Compensation of Victims of Toxic Substance Pollution: A Model Act." *Harvard Journal of Legislation* 14: 683–824.

Soble, Stephen M., and Janis H Brennan. 1988. "A Review of Legal and Policy Issues in Legislating Compensation for Victims of Toxic Substance Pollution." In "Social Policy for Pollution-related Diseases," ed. Michael R. Reich. Special issue of *Social Science and Medicine* 27: 1061–1070.

Sōrifu tōkei kyoku. Various issues, 1960–1982. *Nihon tōkei nenkan* [Japan statistical yearbook]. Tokyo: Nihon tōkei kyōkai and Mainichi shinbun sha.

Starr, Chauncey. 1969. "Social Benefit versus Technological Benefit." *Science* 165: 1232–1238.

——. 1973. "Benefit-Cost Considerations in National Planning." In *Science and Technology Policies: Yesterday, Today, and Tomorrow,* ed. Gabor Strasser and Eugene Simons. Cambridge: Ballinger.

Starr, Chauncey, Richard Rudman, and Chris Whipple. 1976. "Philosophical Basis for Risk Analysis." *Annual Review of Energy* 1: 629–659.

Steiner, Kurt. 1965. *Local Government in Japan*. Stanford: Stanford University Press.

Susskind, L., and J. Cruikshank. 1987. *Breaking the Impasse: Consensual Approaches to Resolving Public Disputes*. New York: Basic Books.

Suzuki, H. 1969. *Hamaoka genshiryoku hatsuden sho mondai ni kansuru shingikai kaichō chōsa* [Deliberative council investigations on problems relating to the Hamaoka nuclear power plant]. Hamaoka: Hamaoka genshiryoku hatsuden sho mondai taisaku shingikai.

——. 1977. *Hamaoka genshiryoku hatsuden to watashi tachi no kurashi: sono anzen sei to kangae kata* [The Hamaoka nuclear power plant and our lives: Safety and a way of approaching safety issues]. Tokyo: Denryoku shinpōsha.

Tanaka, K. 1981. "Genpatsu richi to chiiki shakai hokkaido: Iwanai" [Nuclear siting and regional society: Hokkaido] In *Chiiki funsō no kenkyū: Jichi tai no yakuwari to gōi keisei no jōken* [Research on regional conflict: The role of local administration and conditions for public acceptance], ed. T. Inuta and K. Nagatani. Tokyo: Gakuyō shōbō.

Touraine, Alain, et al. 1983. *Anti-Nuclear Protest: The Opposition to Nuclear Power in France*. Cambridge: Cambridge University Press.

Toyoda, M. 1976. *Genshiryoku hatsuden gijutsu dokuhon* [All you needed to know about nuclear generating technology]. Tokyo: Ōmusha.

Tsūshō sangyō shō. Various issues, 1960–1982. *Denki jigyō binran* [The electric power industry handbook]. Tokyo: Denki jigyō rengō kai tōkei iinkai.

———. 1963a. *Dengen kaihatsu tō ni tomonau sonshitsu hoshō kijun* [Compensation standards accompanying electric power developments]. Tokyo.

———. 1963b. *Dengen kaihatsu tō ni tomonau sonshitsu hoshō saisoku* [Compensation guidelines accompanying electric power developments]. Tokyo.

———. 1982. *Dengen sanpō kankei horei shu* [A collection of laws governing the application of the three laws]. Tokyo: Tsūshō sangyō shō.

Upham, Frank K. 1987. *Law and Social Change in Postwar Japan.* Cambridge: Harvard University Press.

Van Wolferen, Karel. 1989. *The Enigma of Japanese Power: People and Politics in a Stateless Nation.* London: MacMillan.

Von Meter, Donald, and Carl Von Horn. 1975. "The Policy Implementation Process: A Conceptual Framework." *Administration and Society* 6 (February): 445–488.

Weinberg, Alvin M. 1972. "Science and Trans-Science." *Minerva* 10, 2: 209–222.

Williams, David. 1985. "Beyond Political Economy: A Critique of Issues Raised in Chalmers Johnson's MITI and the Japanese Miracle." *East Asia: Internaional Review of Economic, Political, and Social Development.* Vol. 3, Frankfurt: Campus.

Yomiuri shinbun [Yomiuri newspaper].

Yomiuri shinbun: hokkaido ban [Yomiuri newspaper: Hokkaido edition].

Index

Abandoned nuclear projects, 2, 42, 49–51, 61, 79, 148
Acronyms for siting problems, 1, 8*n*, 108, 148–149
Agency for Natural Resources and Energy. *See* ANRE
Ainu, 124–125
Allison, Gary, 7, 9*n*
Amalgamation, 68–69, 89
ANRE (Agency for Natural Resources and Energy), 109. *See also* MITI
Aoki, Masahiko, 7*n*, 9*n*
Apter, David, 3, 7, 10
Aqua, Ronald, 143*n*
Auctioning strategies, 63, 67. *See also* Strategies

Bacow, Lawrence, 10
Bargaining, 3–4, 11–12, 135. *See also* Institutional power
    best alternative to a negotiated agreement (BANTA), 9, 139
    as a learning process, 12, 98
    and limits to, 13–14
    and noncompliance, 35
    and power, 78, 97, 115, 133–134, 139, 145–146
Baumol, William, 143*n*
Belgium, 38
Broadbent, Jeffrey, 2*n*, 7

Calder, Kent, 7*n*, 9*n*, 18*n*, 10, 43, 139*n*, 140
Canada, 8, 38, 139
Case studies, 16–17, 145–148
    choice of, 53–56
    objectives of, 56–59
CBA (Cost-Benefit Analysis), 19, 57, 152
Chugoku Electric Power Company, 108–109
Coal policy, 100–102, 121–122. *See also* Energy policy
Combined methodologies, 17, 53, 144–145. *See also* Polimetric models; Case studies
Commercial Operating Permit. *See* COP
Community, 12
    and attitudes toward environment, 47, 106, 144, 148, 150–151
Compensation, 9–11, 13, 32–38, 76–77, 91, 98
    as bribery, 76, 95, 138

183

and cooperation money, 34, 74, 131–132
cross-national comparisons, 37–38, 141, 153
and risk mitigation, 10, 74, 112, 127–131
and subsidies, 72, 108–109, 128
and Three Laws, 90, 111–112, 126
trends, 149–150
use as a policy tool, 10, 32–33, 139–141
Condron, Margaret, 6, 13
Conflict, 4–7, 10–13. *See also* Bargaining; Spillover effects
intensity of, 1–2, 136–137
Construction Planning Permit. *See* CPP
COP (Commercial Operating Permit), 25–28
Cordes, Joseph, 152–153
Cost-Benefit Analysis. See CBA
CPP (Construction Planning Permit), 25–27
Creatively differentiated models, 18–19, 151–152
Crisis and compensation, 10, 140
Culture, 6, 19–20, 153

Date oil-fired plant, 124–125
Delay, 1–2. *See also* Lead time
Demonstration effect, 63, 85, 89–90
Donnelly, Michael, 9

Easterling, Douglas, 10, 140
ECCS (Emergency Central Cooling System), 121
Economic Planning Agency. *See* EPA
EIA (Environmental Impact Assessment), 27, 43, 105–107, 110–111. *See also* Marine investigations
Electric Power Development Company. *See* EPDC
Electric Power Development Coordination Council. *See* EPDCC
Electric power industry, 22
Electric Utility Industry Law, 22
revisions to, 150
Electricity distribution networks, 22, 71, 81, 102–103, 108–109, 118
Electricity market forecasts, 22, 44–45
accuracy of, 57–60

and incentives to site, 62, 71, 108, 118, 124–126
Electricity market spheres, 25–26, 44–45
Electricity prices, 36, 103
Emergency Central Cooling System. *See* ECCS
Eminent domain, 32, 38, 113, 153
Energy policy, 3, 38–39, 104–105, 125–126, 151
and Three Laws, 35–37
Environmental Impact Assessment. *See* EIA
Environmental quality, demand for, 46, 142–143
EPA (Economic Planning Agency), 25
Elections, 83, 123–124, 128. *See also* Recall movements; Referendum
EPDC (Electric Power Development Company), 22
role in energy policy, 100, 104–105
and site selection, 100–101
EPDCC (Electric Power Development Coordination Council), 25, 45, 95, 105, 113, 131–132
Europe, 20, 31, 137, 153
Expectations, 15, 78–79, 98, 116, 134, 147
Externalities, 12. *See also* Spillover effects

Fisher, Roger, 9
Fishing cooperatives, 22–23
and industry structure, 65–66, 87–88, 107, 120
France, 8, 20, 36, 38, 139–141
Friedman, David, 7*n*
Fukui, Haruhiro, 7*n*, 18*n*

Germany, 20

Haley, John, 9*n*
Harootunian, H., 146*n*
Hata, Toju, 85–88
ousting as fishing leader, 93–96
Hiroshima, 23
Hydroelectric plants, 22, 36, 81

Ideology, 47, 143. *See also* Inter-party conflict
Income derived payment formula, 34, 91. *See also* Compensation

Independent Power Producers. *See* IPP
Inoguchi, Takashi, 7
Institutional power, 9, 138–139, 151
Interest groups, 14, 145, 149–150. *See also* Spillover effects
Inter-party conflict, 47, 84, 87, 107, 123–124
Intra-party conflict, 66–70, 75–76, 83, 97, 122
IPP (Independent Power Producers), 150–151
Ishikawa, Maruyoshi, 66
Iwanai fishing cooperative, 120–121, 126, 130–132

Johnson, Chalmers, 7

Kabashima, Ikuo, 7
Kamogawa, Tadaichi, 82–83, 91–92
Kamoura, Fusahiko, 111, 113
Kansai Electric Power Company, 62
Kato, Osaburo, 81–82
Kawaguchi, Yuzo, 94–95
Kawarazaki, Mitsugi, 83–84
Kowaura fishing cooperative, 64–66
Krauss, Ellis, 2n, 7, 9, 139, 148n
Kubo, Tadaishi, 101–102
Kubokawa, Aiikichi, 92–93
Kunreuther, Howard, 10, 36, 140
Kurebayashi, Matsutaro, 92

Lead time, 21, 29–31, 148
 and measurement difficulties, 25, 28
 trends, 53, 149
Least-cost methods, 118–120. *See also* CBA
Lewis, Jack, 3, 143n
Licences for siting, 27–28
Lipsky, Michael, 6
Local Financial Index (LFI), 44, 46
Locational features, 22–33, 150–151

Magorian, Christopher, 143
Maki nuclear plant, 17–18, 149
Marine investigations, 72, 105–106, 109, 127–128
Maruo, Kenji, 82

Matsushima fishing cooperative, 107, 112–113
Matsushima Kosan, 101, 111
McKean, Margaret, 2, 7, 10, 47n, 143n, 148n
Meiji Fishing Laws, 33
Ministry of Finance. *See* MOF
Ministry of International Trade and Industry. *See* MITI
MITI (Ministry of International Trade and Industry), 7, 22, 45
 and compensation mechanisms, 33–37
 intervention by, 71, 74–76, 109–110, 130
 and role in siting, 32, 137–138
Mizuno, Shigeru, 82
MOF (Ministry of Finance), 101, 104–105, 109–113
Morell, David, 143
Mouer, Ross, 9n
Muramatsu, Michio, 7

Nagahama, Toshio, 122–124, 128
Nagasaki, 23
Nagashima Incident, 75
Nagata, Kenji, 106
Nakamoto, Ichiro, 95
Nakaseko, Bunji, 69–70
Nakasone, Yasuhiro, 74–75
Nakatani, Iwao, 7n
Narita International Airport, 3, 10, 129
Nomura, Junnosuke, 63, 66–67
Nuclear accidents, 121, 140
Nuclear safety, 10, 48, 64, 126, 140
 and earthquakes, 84–85
 and terrorist threat, 129–130
Numazu oil project, 93

Oates, Wallace, 143n
O'Hare, Michael, 10, 154
Okimoto, Daniel, 7n, 9n
Omaezaki fishing cooperative, 89, 94
Ono, Yasuhiro, 85, 87–88
Onoda, Shosaku, 95
Onogawa nuclear plant, 29, 127

Pearl cultivation, 65
Pempel, T. J., 4, 7n, 18n
Petition, 73

Pharr, Susan, 9n, 10, 139n, 140
Polimetric models, 16–17, 41–49, 77, 142–144
  assumptions of, 57–59, 144–145
  and use as screening device, 152
  utility of, 49–51, 57–59
Politics of reciprocal consent, 3, 22, 139
Popper, K., 1n
Portney, Kent, 10, 141
Potential Pareto improvement, 19
Power plant categories, 28
  and ease of siting, 29–31, 148–150
Preemption, 8, 113, 153
Prestowitz, Clyde, 18n
Primary industry, 47, 144–145. See also Property rights
Project impacts, 52, 148–149
Property rights, 33–35, 52
  and agricultural cooperatives, 120, 90–92
  and fishing cooperatives, 64, 74, 88–89, 113, 120, 131–132
  and individuals, 111
Protest as a political resource, 6, 148. See also Resource mobilization models

Ramseyer, Mark, 7n
Rationality of opposition, 6
Recall movements, 67, 76, 149
Reed, Steven, 7
Referendum, 18, 149
Regression models, 16–17. See also Polimetric models
Resource mobilization models, 6
Reverse NIMBY syndrome, 51–53, 107–108, 131
Risk, 48. See also Nuclear safety
  and benefit trade-offs, 10, 141
  familiarity with, 48, 52, 107
Rohen, Thomas, 9n, 139
Rosenbluth, Frances, 7n

Sagara fishing cooperative, 85, 95
Sakaguchi, Yuzo, 68–70
Sakura agricultural cooperative, 82, 91–92
Samuels, Richard, 3–4, 7n, 8, 9n, 10, 22, 139, 143

Sanderson, Debra, 10
Sasaki fishing cooperative, 69–70, 74
Satonaka, Masahira, 66
Sawa, Nagayo, 3, 7, 10
Sheard, Paul, 7n
Shikoku Electric Power Company, 108–109
Shinkansen, 47
Silberman, Bernard, 146n
Simcock, L., 2n, 148n
Similar situation derived payment formula, 34, 91, 127, 131. See also Compensation
Sipher, Dixie, 10
Site selection criteria, 22–23, 62–63, 74, 81–83, 118–120, 127, 130–131.
Sone, Yasunori, 7, 9n
Spain, 20, 38
Spillover effects, 7, 12–14, 42–43, 115, 133
  on communities, 64–65, 77, 97, 142–145
  on utilities, 71, 102–104, 127
State theory, 7–8, 138–139
Steiner, Kurt, 7n
Steinhoff, Patricia, 9n, 139
Steps in siting, 23–28
Strategies, 10, 15, 78, 97–98, 115–116, 134, 146
  coalition, 64–66, 85–88, 120–122
  marginalizing resistance, 72–73, 92–95, 108–109
  pressure, 74–76, 109–110, 130, 94–95
  scheduling negotiations, 91, 111–112, 124–125, 131–132
  site relocation, 76, 130–131
Sugimoto, Yoshio, 9n

Tanaka, Satoshi, 62–63, 66, 73
Taniguchi, Tomomi, 69, 74–76
Technology development, 62, 118
Tohoku Electric Power Company, 29
Tokai nuclear plant, 63
Tokyo bridge, 3, 8, 10
Tomari fishing cooperative, 120–121, 131

Uncertainty, 15, 79, 98, 116, 134, 147
United States, 8, 22, 31, 36–38
  as a comparative referent, 4, 20, 139–141, 153–154

Upham, Frank, 9n
Ury, William, 9

Van Wolferen, Karel, 9n, 18n

Weisbrod, Burton, 152–153

Yamamoto, Tenzo, 67
Yanagihara, Seiji, 94–95
Yoshida, Tamenari, 63, 68–69, 74–76